Body and Brain

DALE PURVES

Body and Brain

A Trophic Theory of Neural Connections

Harvard University Press
Cambridge, Massachusetts, and London, England

Printed in the United States of America
10 9 8 7 6 5 4 3 2

First Harvard University Press paperback edition, 1990

Library of Congress Cataloging in Publication Data

Purves, Dale.
Body and brain.

Bibliography: p.
Includes index.
1. Neural circuitry. 2. Nervous system—Growth.
3. Neuroplasticity. 4. Developmental neurology.
I. Title. [DNLM: 1. Neurons—physiology. WL 102.5 P986b]
QP363.P87 1988 599'.0188 88-764
ISBN 0-674-07715-6 (alk. paper) (cloth)
ISBN 0-674-07716-4 (paper)

Contents

1 Introduction 1

Definitions 1

Historical Background 4

The Trophic Theory as an Organizing Principle 14

2 Effects of Animal Size and Form on the Organization of the Nervous System 17

Somatic Representation 17

Influence of Size on the Nervous System 19

Influence of Form on the Nervous System 27

Significance of Neural Adjustment to Changes in Size and Form 33

3 Coordination of Neuronal Number and Target Size 35

Neuronal Populations in Small, Simple Animals 35

Neuronal Populations in Large, Complex Animals 38

Mechanisms That Coordinate Neuronal Number and Target Size 44

Some Provisos 51

Regulation of Neuronal Populations by Their Targets in Ontogeny and Phylogeny 52

4 Neuronal Form and Its Consequences 54

Diversity of Neuronal Form 54

Neuronal Form in Small, Simple Animals 55

Neuronal Form in Large, Complex Animals 59

Neuronal Form in Phylogeny 64

Functional Consequences of Neuronal Geometry 67

5 Regulation of Developing Neural Connections 75

Formation of Neuronal Connections in Vertebrates 75

Synaptic Rearrangement in the Peripheral Nervous System 77

Competitive Nature of Synaptic Rearrangement 81

Synaptic Rearrangement in the Central Nervous System 87

Purposes of Synaptic Rearrangement 95

6 Regulation of Neural Connections in Maturity 97

Target-Dependent Neuronal Survival 98
Experimentally Induced Changes of Connectivity 99
Normal Remodeling of Axonal and Dendritic Branches 107
Uncertainties about Plasticity of Central Connections 118
Reasons for Ongoing Trophic Interactions 120

7 A Molecular Basis for Trophic Interactions in Vertebrates 123

NGF as a Regulator of Neuron Survival 123
NGF as a Regulator of Neuronal Processes 128
Effects of NGF on Axons, Dendrites, and Synapses 130
A General Scheme for the Action of Trophic Molecules 134
Other Target-Derived Trophic Molecules 136
Uncertainties about the Biology of NGF 137
The Significance of Trophic Molecules 139

8 Effects of Neural Activity on Target Cells and Their Trophic Properties 142

Effects of Neural Activity on Target Cells 142
Influence of Neural Activity on Retrograde Trophic Signaling 144
Influence of Neural Activity on the Arrangement of Competing Inputs 149
Influence of Neural Activity on Convergent Innervation 151
A Model of Activity-Dependent Trophic Support 153
Some Complications 158
The Significance of Activity-Dependent Modulation of Neural Connections 159

9 Implications of the Trophic Theory of Neural Connections 161

Implications for Different Taxa 161
Implications for Learning and Memory 165
Implications for Regressive Theories of Neural Connectivity 169

10 Conclusion 173

Glossary 177
Bibliography 183
Acknowledgments 225
Index 227

Body and Brain

ONE

Introduction

THIS book explores the proposition that the changing size and form of the bodies of mammals and other vertebrates elicit corresponding changes in the connectivity of the nervous system. For reasons that will become apparent, I refer to this general idea as the trophic theory of neural connections. Because it is conventional to think of the nervous system as an organ that monitors and motivates the body rather than as an organ controlled by the body, the perspective of the trophic theory may appear unusual. Nevertheless, the body's influence on the nervous system is as important for the organism as is neural dominion over the body.

As theories about the nervous system go, the trophic theory of neural connections is decidedly biological, providing no insight into consciousness, intelligence, free will, or other psychological phenomena whose explanation is popularly considered to be the proper objective of a neurological theory. Moreover, there is as yet little evidence that the concepts expressed in the theory apply to animals other than vertebrates, which are, after all, a tiny subset of the animal kingdom. These limitations notwithstanding, the trophic theory is of fundamental importance to topics as diverse as neural development, the response of the nervous system to the inevitable injuries that arise in the course of life, the way in which information is stored in the nervous system, and the manner in which the nervous system adjusts to the somatic changes that occur during the course of evolution.

Definitions

The meaning of most specialized terms used in this book can be found in the glossary. Several key words and phrases, however, need to be defined in detail at the outset. Preeminent among these are the term *trophic* and the general concept of *trophic interaction*.

The word trophic is taken from the Greek *trophē*, which means, roughly, nourishment. Trophic interactions have been operationally defined over the years as long-term dependencies between neurons and the cells they innervate (non-neural cells or other neurons, as the case may be). By long-term, neurobiologists generally mean effects that can be observed over weeks or months; by dependency, they refer to the deleterious effects observed when one cell or the other in the partnership is removed or incapacitated. The indexes of trophic dependence range from neuronal death to a myriad of electrophysiological and metabolic effects taken to be signs of cellular ill health. The agent of these effects is generally envisioned as intercellular signals. In most accounts, these signals are a molecular message that passes between the two cells involved (in addition to whatever neurotransmitters may operate between the cells in question). Historically, the phrase trophic interaction has been applied both to long-term dependencies of nerve cells on their targets and to dependencies of targets on the nerve cells that innervate them.

The sustenance provided to neurons in the course of trophic interactions is not the sort derived from the ordinary metabolites that neurons use to generate the energy and structural elements essential for any cellular activity. Although the agents of trophic interaction may be equally important (neurons deprived of trophic support may die), trophic effects are specific: the requirement of one kind of nerve cell is qualitatively different from that of another. Specific cellular interactions are by no means unique to trophic agents; neurotransmitters and hormones are other classes of molecular signals that influence specific sets of cells in ways that are now well understood.

Some other terms that require definition concern the structure of the nervous system. These include *axon, dendrite, neural target,* and *neural connections.* Axons are usually defined as neuronal processes that traverse a substantial distance and carry a regenerative electrical signal (the action potential) to a distant target. Dendrites are neuronal branches, usually short and profuse compared to axons, that passively receive innervation from other neurons. These conventional definitions are ultimately inadequate because they do not fully describe the complex reality of neuronal processes and their functions. Thus, some neuronal processes which by one criterion appear to be axons also have functional characteristics of dendrites. In other instances, neuronal processes that have the typical appearance of dendrites support action potentials. Indeed, it would be difficult to classify the processes of some neurons. In short, axons and dendrites are usually distinct in both their anatomy and their function; in some cases, however, the differences between them are blurred.

Some nerve cell axons innervate muscles, glands, sensory receptors, and other organs directly. Other neurons innervate the nerve cells that innervate these somatic structures. Still other neurons—the majority—establish their connections entirely within the central nervous system. Thus the *targets* of innervation can be cells outside the nervous system or other nerve cells within it. For the most part, the evidence which supports the trophic theory of neural connections has been drawn from observations of the first-order nerve cells that innervate the body directly and the second-order neurons that innervate or are innervated by the primary motor and sensory neurons. However, every neuron in the vertebrate nervous system is ultimately linked to the body by a chain of neural connections. It is therefore axiomatic that the concepts expressed in the trophic theory apply in some measure to all nerve cells and their connections.

The idea that the body influences the nervous system by affecting the concatenation of neurons linked to somatic structures is discussed in terms of *neural connections*. This phrase refers to the number and disposition of axonal and dendritic branches as well as to the synaptic relationships they establish. The emphasis on the regulation of neural connections admittedly minimizes consideration of other ways in which the body influences the nervous system (for example, by inductive influences on neural differentiation) and other important aspects of synaptic relations (such as the formation of synapses according to intercellular recognition). The short shrift given these issues is intended not to diminish their importance or to obscure the true complexity of the neurosomatic relationship, but simply to maintain a sharp focus on the theme of the changing anatomy of neural connectivity.

In some uses, the phrase *neural connections* refers primarily to synapses, the highly specialized junctions that occur at the points where one nerve cell contacts another. There are good reasons, however, for defining neural connections to include the axonal and dendritic branches that link nerve cells and their targets. One reason is that a good deal of evidence supports the action of trophic signals on neuronal branches. Whether such signals have a *direct* effect on synaptic specializations is for the most part not known. Another reason is that connections between nerve cells and targets do not always involve anatomically discrete synapses. An example is the innervation of viscera by autonomic neurons, in which case the release of transmitter by the presynaptic cells occurs at some distance from the postsynaptic smooth muscle cells (which in turn show no specialization that is confined to the vicinity of the nerve terminal). The innervation of body surfaces by free nerve endings (in the skin and other surface epithelia, for instance) deviates still further from the norm in that no synapses of any type are involved in the association of neuron and target. Trophic action is therefore discussed in terms of effects on axonal and dendritic branches rather than on synapses *per se*.

In the light of these definitions, it is possible to formulate a working definition of the trophic theory itself. The theory holds that patterns of nerve cell connections—which is to say, the number and disposition of axonal and dendritic arbors and the connections they make—are subject to ongoing regulation by interactions with the cells that they contact.

Historical Background

Until ten or fifteen years ago, it was widely supposed that the nervous system is hard-wired, meaning that neural connections form according to a precise plan which remains fixed thereafter, in the style of electrical circuits. There were, and to some extent still are, good reasons for this consensus. The early differentiation of neurons, the paucity of direct evidence for anatomical change in the adult nervous system, and the inability of the mammalian nervous system to compensate effectively for neural injury all argue for a relatively static organization of neural connections. A corollary of this view is that the principal purpose of neural development—with respect to the connections between nerve cells—is to establish such circuits with great accuracy.

The first evidence that nerve cells are endowed with qualities which allow them to generate precise patterns of connectivity was provided by the English physiologist J. N. Langley. In his pioneering work between 1875 and 1925, Langley defined the autonomic nervous system of mammals and many of its properties. The autonomic system is that part of the nervous system primarily concerned with the functions of smooth muscles and glands; as a consequence, it is also referred to as the visceral or involuntary division of the nervous system. The part of this system investigated most thoroughly by Langley was the superior cervical ganglion (Figure 1.1). Neurons in this most rostral of the segmental sympathetic ganglia innervate the gamut of visceral end-organs in the head and neck (the blood vessels, iris, salivary glands, and piloerector muscles, as well as some targets within the brain). Langley

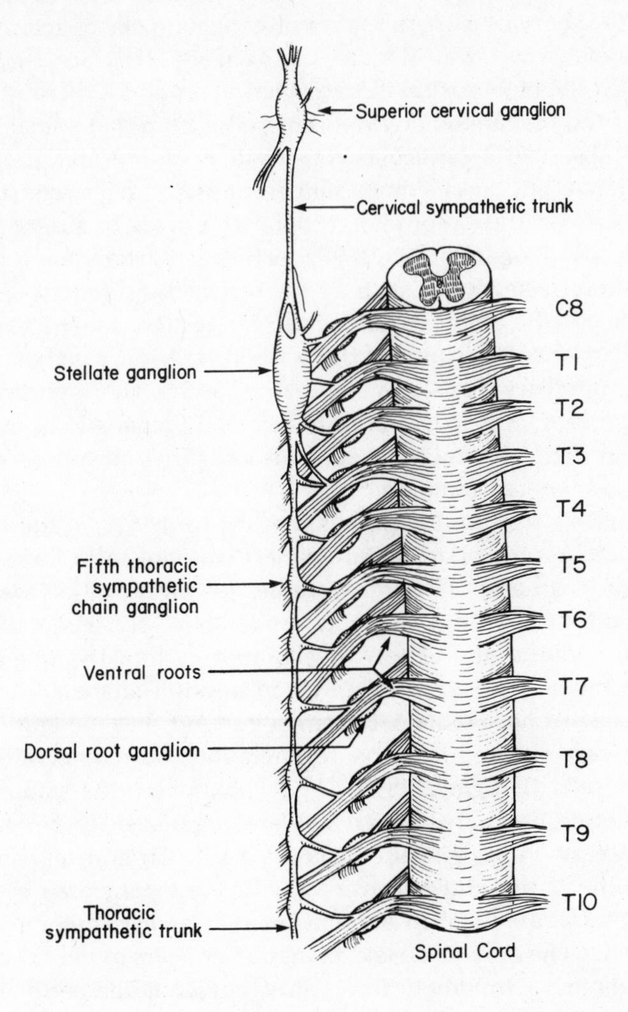

Figure 1.1. Overall arrangement of the cervical and thoracic portion of the sympathetic nervous system in mammals (ignoring minor differences among species). The sympathetic ganglia, each comprising thousands of neurons, receive innervation from preganglionic neurons that reside in the spinal cord; the axons from these spinal cord neurons reach the ganglia by way of the ventral roots. The axons that arise in turn from ganglion cells innervate smooth muscles, glands, and other targets at corresponding segmental levels (indicated on right; C = cervical, T = thoracic). The peripheral part of the sympathetic system is attractive because of both its accessibility and the simplicity of its anatomical organization compared to pathways wholly within the spinal cord and brain.

found that the preganglionic axons arising from different spinal cord segments innervate superior cervical ganglion cells in mammals in a highly stereotyped way (Figure 1.2; Langley, 1892, 1921). When he stimulated the most rostral thoracic ventral root (T1), thereby stimulating all of the preganglionic axons emerging from that spinal segment, Langley observed a particular constellation of end-organ effects: the pupil dilated, but other sympathetic responses, such as constriction of blood vessels in the ear or piloerection, were weak or absent (Langley, 1892; see also Langley, 1895, 1897). Conversely, stimulation of a more caudal thoracic segment, such as T4, revealed a different set of end-organ effects: the blood vessels of the ear became constricted and the hair in part of the territory of the superior cervical ganglion stood on end, but pupillary dilation was weak or absent. These findings led to the supposition of a special affinity between preganglionic axons arising from different spinal cord segments and different subsets or classes of superior cervical ganglion cells.

In Langley's day, an investigation of the formation of these connections in embryonic or neonatal animals was not technically possible (although it is now). Therefore, studies of reinnervation were a reasonable alternative in seeking to understand the general rules of neuronal connectivity. A few weeks after cutting the preganglionic nerve to the superior cervical ganglion in adult animals (the cervical sympathetic trunk; Figure 1.1), Langley found that the end-organ responses were restored by nerve regeneration and reinnervation of the ganglion cells (Langley, 1895, 1897). Moreover, the end-organ responses after nerve regeneration were organized as before. Thus, stimulation of T1 once again elicited a particular constellation of peripheral effects that did not overlap with those generated by stimulation of T4 (Figure 1.2). In summarizing these experiments, Langley concluded: "The only feasible explanation appears to me to be that the [preganglionic] sympathetic fibres grow out along the peripheral piece of nerve . . . spreading amongst the cells of [the] ganglion, and that there is some special chemical relation between each class of nerve fibre and each class of nerve cell, which induces each fibre to grow towards a cell of its own class and there to form its terminal branches" (1895, p. 284).

This straightforward expression of what is now called the chemoaffinity hypothesis was eclipsed for several decades by another conception of how neural connections develop. Why Langley's results and lucid interpretation were ignored for the next fifty years or more is not clear. In any event, a quite different view of neural connectivity was

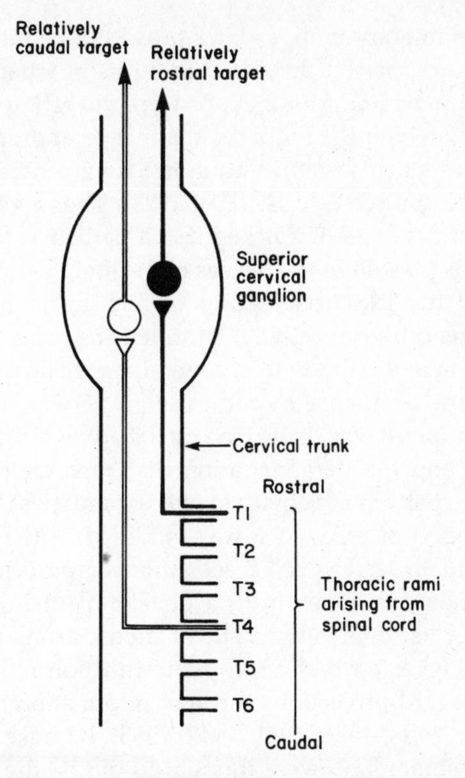

Figure 1.2. Evidence that neural connections form according to specific affinities between different classes of pre- and postsynaptic cells. In the mammalian superior cervical ganglion, preganglionic neurons located in particular spinal cord segments (T1, for example) innervate ganglion cells that project to particular peripheral targets (the eye, for example). The re-establishment of these preferential relationships in adult animals after interruption of the cervical trunk suggests that selective affinities are a major determinant of neural connectivity.

successfully promoted by P. A. Weiss in the context of muscle innervation in vertebrate limbs. In this scheme, put forward at about the time of Langley's death in 1925, specific matching between nerve cells and their targets had an operational rather than an anatomical-chemical basis. The experiments that led to this conclusion involved limb transplantation in amphibians. When an extra limb was placed near the normal appendage in a host animal, the two limbs moved in exact synchrony and with obvious coordination (Weiss, 1924, 1928; see also Weiss, 1936, 1968). Because Weiss could not imagine that the same axons correctly innervated the homologous muscle in both the normal and the supernumerary limb, as later proved to be the case (Stephenson, 1979), he interpreted the result in terms of what would now be called systems matching. Motor axons were thought to ramify more or less indiscriminately in the limb; the manifestly appropriate responses of muscles were taken to represent a matching process based on patterns of neural impulses (Weiss, 1924, 1928). Thus a given muscle was thought to contract when it "picked up" a particular pattern of neural activity that was present in the nerves to the limb. Because this conception embodied the idea that neural targets are somehow tuned to specific patterns of nerve activity, in much the same way that a taut string can be made to vibrate by a tone of a specific frequency, it was referred to as the resonance hypothesis.

When this proposal was disproven by the advent of electrical recording in biology and the demonstration that each motor nerve branch carries a unique pattern of activity to specific muscles (Wiersma, 1931), the original notion of resonance was modified. Although Weiss continued to maintain that specific activation of particular muscles by spinal motor neurons arises from a developmental strategy that involves a largely random outgrowth of motor axons to the limb, he subsequently suggested that appropriate function reflects a reorganization of haphazard projections through information provided by the neural targets (Weiss, 1936, 1941, 1942, 1965). Because the target muscles in the vertebrate limb were thought to dictate the central connectivity of motor neurons that contacted them at random, this revised hypothesis was referred to as myotypic specification. Thus, by the 1930s the groundwork had been laid for a dialectic between the view that the nervous system is wired according to preexisting cellular identities and the alternative belief that neural connections are shaped interactively according to the functional properties of the pre- and postsynaptic elements.

The idea of highly malleable connections capable of being reorganized according to functional criteria was challenged in the late 1930s by R. W. Sperry, a graduate student of Weiss's at the University of Chicago. Sperry revived Langley's original theory in a new context, the retinotectal system, in which axons arising from nerve cells in the retina make an orderly topographic map in the tectal region of the midbrain. He took advantage of the fact that the severed optic nerve regenerates over a period of several weeks in some fish and amphibians, restoring normal vision; indeed, such animals see again even after an eye has been removed and reimplanted, or transplanted from one individual to another (Stone and Zaur, 1940; Stone, 1941; Stone and

Farthing, 1942). To investigate the validity of Weiss's notion of functional plasticity, Sperry cut the optic nerve in newts (and later frogs) and rotated the eye 180° (Figure 1.3; Sperry, 1943a,b; see also Stone, 1944). After the nerve had regenerated, animals with rotated eyes behaved as if their visual world had been inverted and shifted left for right. This outcome—like the outcome of reinnervation in the superior cervical ganglion—suggested that nerve cell axons grew back to approximately the same target cells they contacted originally, in spite of the fact that the regenerated connections in the frog were maladaptive (since they produced nonsensical behavior). This interpretation was confirmed by showing that regenerated fibers, visualized by histochemical staining, appeared to ignore large areas of denervated neural tissue to contact the part of the optic tectum in which they had normally terminated (Attardi and Sperry, 1963; see also Fujisawa, 1981; Fujisawa et al., 1981). Sperry summarized 25 years of work on this issue in the following way: "It seemed a necessary conclusion from these results that the cells and fibers of the brain . . . must carry some kind of individual identification tags, presumably cytochemical in na-

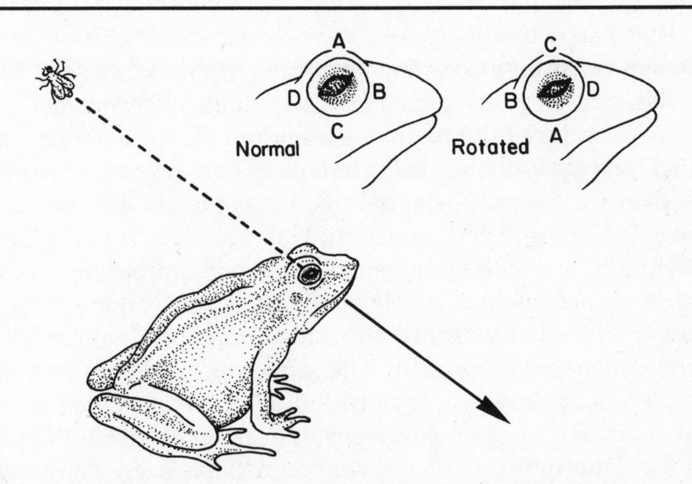

Figure 1.3. Experiment that revived the chemoaffinity theory. The right optic nerve of the frog was cut some weeks earlier, and the eye was rotated through 180 degrees and fixed in place (enlarged above). Although the frog could see perfectly well through the experimental eye, its attack on prey was consistently misdirected. This observation indicates that the regenerating axons of the optic nerve establish connections with their *original* target cells in the brain and that these connections are not reorganized in the face of the functional requirements imposed by new circumstances. (After Sperry, 1956)

ture, by which they are distinguished one from another almost, in many regions, to the level of the single neuron; and further, that the growing fibers are extremely particular when it comes to establishing synaptic connections, each axon linking only with certain neurons to which it becomes selectively attached by specific chemical affinities" (1963, pp. 703–704).

The renaissance of the chemoaffinity theory had a profound effect on the subsequent generation of neurobiologists, leading to a wealth of further experiments after Sperry had retired from the field in the 1960s to concentrate on studies of interhemispheric communication. The hegemony of the chemoaffinity theory, which persists to this day, is justified by the fact that in virtually every system in which this issue has been explored, evidence of selective synaptic connections between pre- and postsynaptic cells has been observed. Nevertheless, important qualifications of the chemoaffinity theory have been required by more recent evidence.

One problem with the theory is the implication that each neuron is unique and can therefore receive only a complementary set of synaptic connections. Although it must in some sense be true that every neuron *is* unique, it is now clear that the distinctions between vertebrate neurons of a given class (for example, tectal neurons) are not terribly rigid. That chemoaffinity is less restrictive than originally imagined was shown by further experiments in the retinotectal system. Thus, if either the retina or the tectum is quantitatively mismatched to its counterpart by surgically removing a portion of the target or a part of the innervating population, adjustments of neural connections are observed over time which belie any rigid preordination of connectivity (Figure 1.4, Gaze and Sharma, 1970; Horder, 1971; Yoon, 1971, 1972, 1976; Schmidt et al., 1978). In general, the retinal projections adjust to occupy the space made available by the particular surgery. This result indicates that retinal neurons can contact target cells other than the ones they innervated originally. The same conclusion has been drawn from experiments in which eyes with duplicate half-retinas have been created at an early stage of development (Gaze et al., 1963, 1965) and in which the pattern of retinotectal connections has been studied during the course of normal development (Chung et al., 1974; Gaze et al., 1974, 1979; Longley, 1978; Meyer, 1978; Easter and Stuermer, 1984; Reh and Constantine-Paton, 1984; Grant and Keating, 1986; O'Rourke and Fraser, 1986). In the case of surgically constructed compound eyes, the axons from both retinal halves project to the entire tectum. Since the experimental eyes comprise duplicate half-retinas, this result again im-

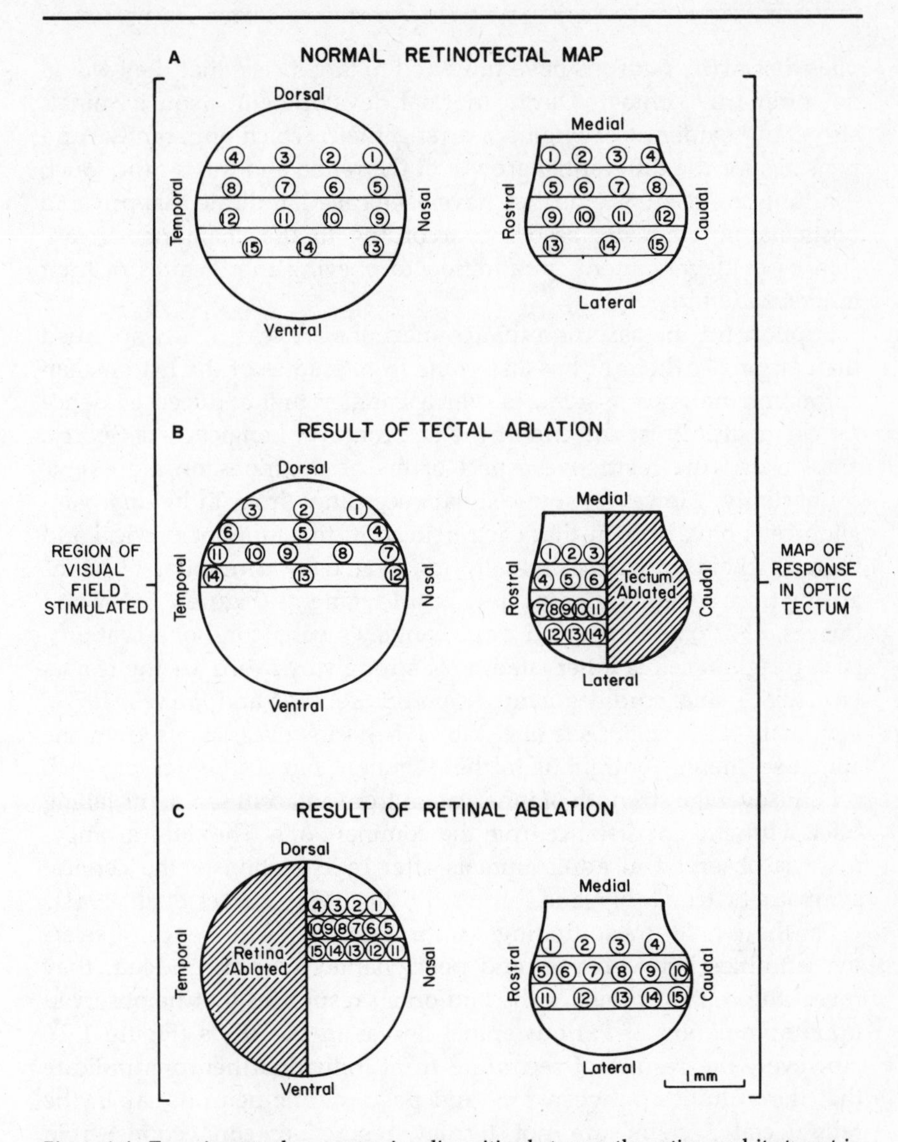

Figure 1.4. Experiments that create size disparities between the retina and its target in the midbrain, the optic tectum, used to test the rigidity of matching during synaptogenesis. (*A*) The normal retinotectal map in a goldfish. The numbers indicate the region of the tectum activated by stimulation of the corresponding point in the visual field. (*B*) A complete, but compressed, retinotectal map several months after the optic nerve was cut and about half of the optic tectum removed. Over time, axons arising from the retina reorder their connections so that the remaining tectal territory is divided among them, in spite of the fact that this entails many abnormal connections. (C) The result of a complementary experiment in which half of the retina was removed at the time the optic nerve was cut. The remaining part of the retina maps over the full tectum. Such compensatory changes suggest that retinotectal connections form according to preferences rather than rigid restrictions. (After Yoon, 1971; Schmidt et al., 1978)

plies that many neurons have projected to target cells that they would not ordinarily contact. During normal development, axon terminals also show evidence of onging rearrangement which apparently compensates for the differential growth of the retina and the tectum. Such "shifting connections," as they have been called, indicate that pre- and postsynaptic elements associate according to the changing circumstances of development, in addition to obeying the dictates of their inherent identities.

Support for the assertion that connections are less rigidly specified than originally thought has also come from studies of the mammalian autonomic nervous system, in which Langley first adduced evidence for chemoaffinity at the end of the last century. Langley, like Sperry, emphasized the restrictive aspect of his observations on autonomic connectivity. However, intracellular recordings from individual ganglion cells have shown that each neuron in the superior cervical and thoracic chain ganglia is actually innervated by a number of axons arising from several different spinal cord segments (Figure 1.5; Njå and Purves, 1977a). As expected from Langley's behavioral observations, each ganglion cell is innervated most strongly by axons arising from a particular spinal cord segment within this set (Njå and Purves, 1977a; Lichtman et al., 1980; see also Yip, 1986). However, axons from the spinal segments contiguous to the dominant one also innervate each cell, the average strength of innervation from adjacent segments falling off as a function of distance from the dominant one. The same arrangement is observed in adult animals after regeneration of the cervical sympathetic trunk (Njå and Purves, 1977b, 1978b; Purves et al., 1981).

On the whole, these findings confirm Langley's inference of selective affinities between pre- and postsynaptic neurons; indeed, they reveal the cellular basis for the end-organ responses that he observed upon stimulation of various spinal nerves in the 1890s (Figure 1.2). However, the results of recording from individual neurons indicate that the affinities between pre- and postsynaptic neurons, as in the retinotectal system, are not terribly restrictive; connections from neurons of a particular spinal level are preferred, but terminals arising from neurons at other levels are not excluded. This interpretation has been further strengthened by quantitative mismatch experiments, similar in principle to those carried out in the retinotectal system. Thus, if a portion of the innervation to the superior cervical ganglion is surgically removed (analogous to experiments in the retinotectal system in which part of the retina is removed), then recordings from individual ganglion cells indicate that new connections are established

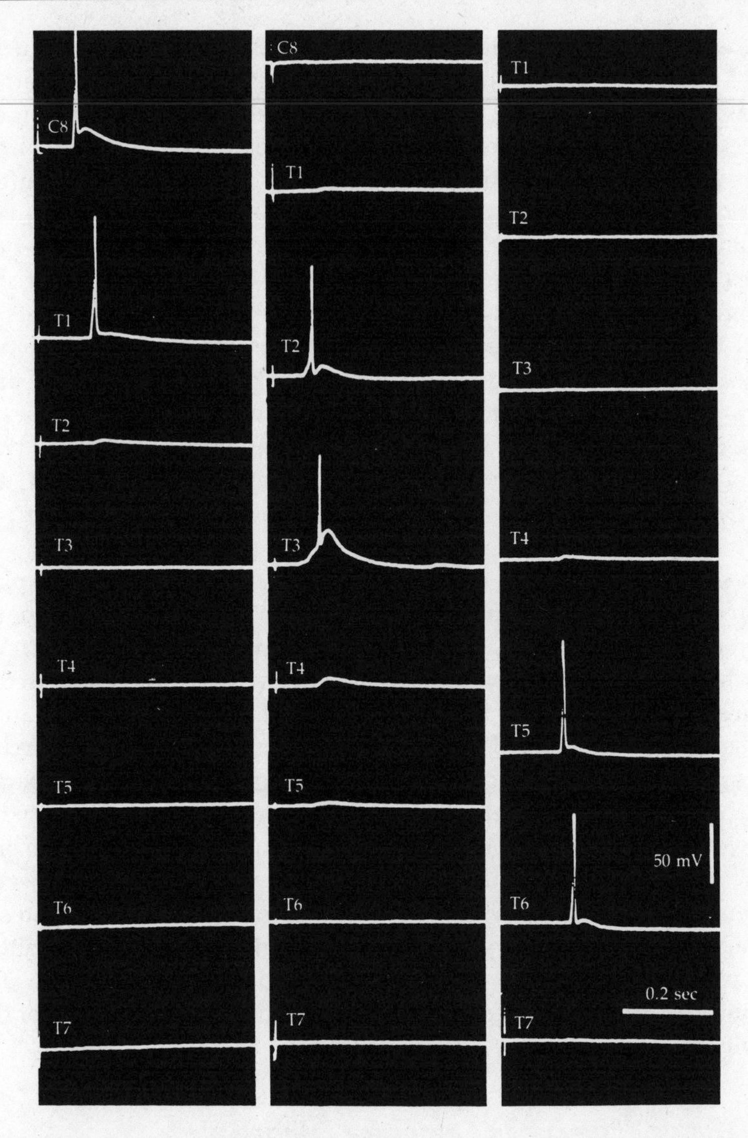

Figure 1.5. Electrical recordings from individual autonomic ganglion cells, which support the view that the affinities operating during the formation of synaptic connections in vertebrates represent biases rather than rigid restrictions. The three vertical panels are intracellular records from three different nerve cells in the superior cervical ganglion of a guinea pig during stimulation of the ventral roots indicated. Each neuron receives strong innervation (indicated by responses which reach threshold and produce an action potential) from one or two spinal segments, with the strength of the synaptic contribution from adjacent segments falling off as a function of distance from the dominant segment. This arrangement, which is re-established during reinnervation, suggests the operation of continuously graded preferences between pre- and postsynaptic cells in this part of the nervous system. (After Njå and Purves, 1977a)

by the residual axons (Maehlen and Njå, 1981, 1984; see also Murray and Thompson, 1957). The novel connections also establish a pattern of segmental preferences, as if the system attempts to do the best it can under the altered circumstances. From a broader vantage, of course, there *are* absolute restrictions to synaptic associations. Thus, neurons do not innervate nearby glial or connective tissue cells, and many instances have been described in which various nerve and target cell types show little or no inclination to establish innervation. When synaptogenesis does proceed, however, neurons and their targets in both the central and the peripheral nervous systems of vertebrates appear to associate according to a continuously variable system of preferences. Such biases guide the pattern of innervation that arises in development or reinnervation without limiting it in any absolute way. The target cells in these examples are certainly not equivalent, but neither are they unique with respect to the innervation they can receive.

In short, the notion of chemoaffinity cannot, by itself, fully explain the patterns of connections in the nervous systems of vertebrates or the manner in which these patterns arise. Thus, neither the view of Weiss, which held that neural connections are governed by functional circumstances, nor the view of Sperry, which held that connectivity can be explained by a rigid set of chemoaffinities, has been fully supported by the evidence accumulated during the last two decades. The apparent simplicity of an experiment or the force of a particular personality has allowed one or the other of these extremes to hold sway for a surprisingly long time. The truth, however, lies somewhere in between. Whereas the nervous system of mammals and other vertebrates is certainly prefigured in a plan that involves inherent cellular affinities, the plan is not decisive at the level of individual nerve cells. The trophic theory of neural connections has emerged in the context of this ambiguity.

The Trophic Theory as an Organizing Principle

The argument for the utility of the trophic theory as an organizing principle begins with the demands that the changing size and form of animals—both in ontogeny and in phylogeny—place on the nervous system. As the size and form of an animal change, whether in development or in speciation, the nervous system must change accordingly in order to preserve efficient function. Thus, neurons could be added to (or subtracted from) the nervous system in proportion to somatic change, or the same number of neurons could make more (or fewer) connections with the relevant targets. In fact, vertebrates make use of both of these strategies.

Neuroembryologists have understood for several decades that the number of neurons in developing vertebrates is influenced by cellular mechanisms that depend on feedback from neural targets and consequent adjustment. The currently accepted view of such regulation, buttressed by a great deal of experimental work, is that targets elaborate trophic signals to which neurons are selectively sensitive. Nerve cells that innervate a particular target must acquire a share of trophic support or die. As a result, the number of neurons that survive to maturity is not fully specified in advance; rather, neuronal populations are defined by the needs of their targets.

The trophic theory extends the general concept of trophic interactions from neuron *survival* to neuronal *connectivity*, that is, to the regulation of axonal and dendritic branches. One reason for broadening the context of trophic action in this way is a growing appreciation among neurobiologists of the general importance of neuronal form and its regulation. Another reason is the increasing evidence that patterns of neural connections are influenced interactively by feedback from neural targets. Like the regulation of number, the governance of form is competitive: the number and disposition of neuronal branches are determined in the course of events that involve peers striving for success vis-à-vis the same target.

The significance of the trophic regulation of neuronal connections is underscored by the further realization that patterns of connections are maintained in maturity by *ongoing* interactions with target cells. This conclusion is based on both experimental interference with established patterns of connections and direct observation of ongoing change in the patterns of connections in the nervous systems of mature animals. The fact that experimental perturbation can alter patterns of connections in maturity demonstrates a persistent potential for change. The additional observation that connections between nerve cells and their targets are subject to ongoing rearrangement indicates that this potential is often realized in normal circumstances.

The legitimacy of trophic interaction as a principle of neural organization—and the usefulness of this perspective in rationalizing a range of seemingly diverse neural phenomena—does not stand or fall on the vindication of any particular cellular or molecular mechanism. A variety of scenarios could explain how neurons receive feedback from the cells they contact and adjust their connections accordingly. Nevertheless, at least one well-studied molecule, the protein nerve growth factor (NGF), is clearly a target-derived trophic agent for several classes of vertebrate nerve cells. The action of NGF, as presently understood, suggests a molecular paradigm for the manner in which neural targets influence the connections of the innervating population.

This detailed knowledge of trophic molecules derived from targets and their retrograde effects tends to obscure the importance of oppositely directed trophic effects. Although the basis of anterograde trophic action is not so well understood, anterograde trophic effects are mediated at least in part by the electrical activity of the cells involved. Many properties of both neural and non-neural target cells, including the production of target-derived trophic signals, are affected by the level of activity generated by neurotransmission over the long term. As a consequence, a regulatory loop is established in which retrograde and anterograde influences can interact; neurons are informed about the state of innervation of their targets, and targets are informed about the success or failure of their efforts to attract and maintain innervation.

The chapters that follow explore the evidence for, the uses of, and the implications of these ideas about neuronal connectivity during development and in adult animals. Neurobiologists tend to suppose that the primary purpose of change in the nervous system is to encode experience. However, neural adjustment is necessary for at least one other reason: the body, as well as the external environment, changes. This transfiguration is most obvious during embryonic development, but it is readily apparent in mammals and other vertebrates throughout a long period of maturation, as well as in the changes that occur in bodies across generations as animals evolve. In order to monitor and motivate a body that is changing in both size and form, the nervous system must also change.

In the interest of maintaining attention on these concepts, this book does not attempt to give a complete account of all the factors that are involved in synaptogenesis and its control. It indicates what these omissions are, however, and calls attention to various weaknesses and uncertainties in the account, of which there are many. This is, after all, the description of a *theory*, a word that is used here in its primary sense of a contemplation of what might be the case. Should the theory be misguided in its particulars, it may nonetheless stimulate a fuller and more accurate consideration of one of the major strategies that the vertebrate nervous system employs to respond to the ever-changing circumstances in which animals live.

TWO

Effects of Animal Size and Form on the Organization of the Nervous System

THE SIZE and form of animals are rarely mentioned in the neurological literature, or even in broader biological contexts in which the dimensions of animals are bound to be important. J. B. S. Haldane expressed the surprise one feels at this rather routine omission in his essay "On Being the Right Size": "The most obvious differences between different animals are differences of size, but for some reason the zoologists have paid singularly little attention to them. In a large textbook of zoology before me I find no indication that the eagle is larger than the sparrow, or the hippopotamus bigger than the hare, though some grudging admissions are made in the case of the mouse and the whale. But yet it is easy to show that a hare could not be as large as a hippopotamus, or a whale as small as a herring . . . and a large change in size inevitably carries with it a change of form" (1985, p. 1).

An argument for the importance of trophic relationships in neural organization properly begins with a consideration of the ways in which changes in the body are reflected in the nervous system. The most obvious of these changes—and certainly the easiest ones to observe and measure—are those of size and form as animals grow (or evolve). A number of basic facts about ontogeny and phylogeny indicate that the nervous system must adjust continually to the size and form of the body that it controls.

Somatic Representation

Fundamental to the trophic theory is the idea that the body is repeatedly represented within the nervous system (Figure 2.1). The existence of such neural maps was first suggested by the clinical observation that the abnormal bodily sensations experienced in certain kinds of epileptic seizures (called Jacksonian fits) progress in a characteristic sequence (for example, hand, forearm, upper arm, shoulder). This

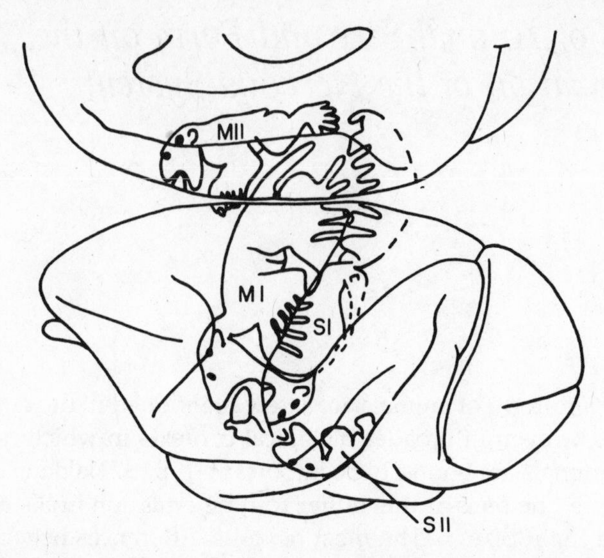

Figure 2.1. Cerebral cortex of the Macaque monkey, showing the arrangement of several maps (M I and II and S I and II indicate the primary and secondary motor and sensory maps). The left hemisphere is shown with the medial portion swung upward to reveal the portions of the maps that are on the medial surface of the hemisphere. The motor maps were determined by observing the part of the body that responds to focal stimulation of the cortex; the sensory maps were assessed by recording locally evoked potentials in the cortex in response to natural stimulation (touch, for example) of different regions of the body surface. (After Woolsey, 1958)

paresthetic march implies the spread of an electrical disturbance within an orderly central representation or "schema" of the body (Sittig, 1925; see also Fritsch and Hitzig, 1870). With the advent of electrical stimulation and recording techniques in the 1930s, neurosurgeons operating on epileptic patients in order to remove a seizure focus confirmed the existence of such schemata and extended the concept of maps to motor as well as sensory representations (Penfield and Boldrey, 1937; Penfield and Jasper, 1954). Such observations were made by stimulating and recording from the cerebral surfaces of these individuals under local anesthesia (because the brain itself has no sensation, neurosurgery is often carried out under these conditions). More detailed experimental studies in monkeys showed that the body is indeed represented in a systematic manner in both the somatosensory and the motor cortices (Bard, 1938; Woolsey et al., 1942; Woolsey, 1958).

The developmental processes that lead to orderly central maps are clearly complex. First, maps are preserved across the synaptic linkages (often several) between the body and the central nervous system. Second, motor and sensory maps reflect the relative importance of various parts of the body, rather than simply body geometry. Thus, somatic structures that are densely innervated by sensory endings and used for delicate movements (the fingers and lips, for instance) occupy a disproportionately large part of somatic maps. Third, the maps are in some instances discontinuous. Finally, there are multiple motor and sensory maps of the body. In spite of these and other complications, the existence of neurosomatic maps makes a fundamental point: changes in the body must be accompanied by changes in the neural representation of the body. By the same token, there must be interactive mechanisms that allow these adjustments to occur.

Influence of Size on the Nervous System

Perhaps the most obvious somatic change is a change of size. It tends to be taken for granted that the adult version of most species is much larger than the fully formed embryo. The growth of mammals, from the late fetal stages to the adult, is always substantial and in many cases astounding (Figure 2.2). During the first few months of postnatal life, a blue whale, the largest extant mammal, gains weight at the prodigious rate of 10^5 grams per week (Small, 1971). Adult blue whales weigh on the order of 10^8 grams (such estimates are considered conservative, because whales are weighed piecemeal, after blood and other body fluids have been lost; see Mackintosh and Wheeler, 1929); in contrast, a fetus 50 centimeters long weighs only a few thousand grams. Differences in animal size across species are equally striking.

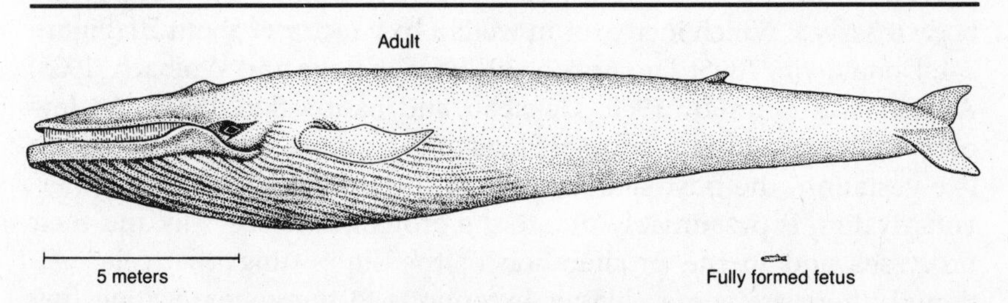

Figure 2.2. Size of an adult blue whale (*Balaenoptera musculus*) compared to size of the fully formed fetus. The adult is a female 25 meters in length; the fully formed fetus is about 50 centimeters long. The drawing is approximately to scale. (After Mackintosh and Wheeler, 1929)

The smallest mammal, the common shrew, weighs about 4 grams. Differences in body weight between various mammals may therefore be as much as 50-million-fold (Calder, 1984; Schmidt-Nielsen, 1984).

These impressive statistics raise an obvious question about the nervous system that is pertinent to both development and evolution. How does this organ manage to initiate coordinated movement in, and monitor the sensations of, bodies that differ so markedly in bulk?

In order to answer this question in the context of development, it is necessary to know some further facts about the growth of the body vis-à-vis the growth of the nervous system. The postnatal growth of vertebrates occurs in two general ways: cell addition and cell enlargement. Most cell types (for example, cells of the gut, skin, liver, blood, and many endocrine glands) continue to divide during maturation; thus the growth of the relevant organs or tissues occurs primarily by the generation of new cells (Levi, 1905; Teissier, 1939; Altman and Dittmer, 1962; see also Conklin, 1911). Other cell types (skeletal and cardiac muscle, for example) are generated in embryonic life and do not usually divide thereafter; accordingly, the growth of muscles during the postembryonic maturation of vertebrates usually (but not always) involves the enlargement of existing cells.

The nerve cells of mammals also fall into this latter category in that the division of neuroblasts ceases relatively early, usually by late embryonic stages (Figure 2.3; Angevine and Sidman, 1962; Altman, 1963; Altman and Das, 1965; Angevine, 1965; Rakic, 1974; Sidman and Rakic, 1982; Rakic, 1985a,b). In man, one indication of this fact is the disproportionately large size of the brain at birth. The brain of a newborn (male) infant weighs about 350 grams, on average; that of an adult man typically weighs about 1400 grams. Since the average weight of a newborn is 3,400 grams and of an adult is 70,000 grams, the mass of the nervous system increases far less (about 4-fold) than the mass of the body it serves, which increases in weight by a factor of about 20 (Figure 2.4; Donaldson, 1895; Thompson, 1917; Coppoletta and Wolbach, 1933; Altman and Dittmer, 1962; Dekaban and Sadowsky, 1978). As few nerve cells are added to most regions of the primate brain after mid to late gestation, the postnatal increase in the weight of the human nervous system is presumably due to the growth of nerve cells and their processes and to the proliferation of the supporting glial cells. Although there are some obvious exceptions to the generalization that all the neurons which will be present in adult mammals are generated in the embryo (Altman, 1963; Altman and Das, 1965; Altman and Bayer, 1977; Altman and Winfree, 1977; Kaplan and Hinds, 1977;

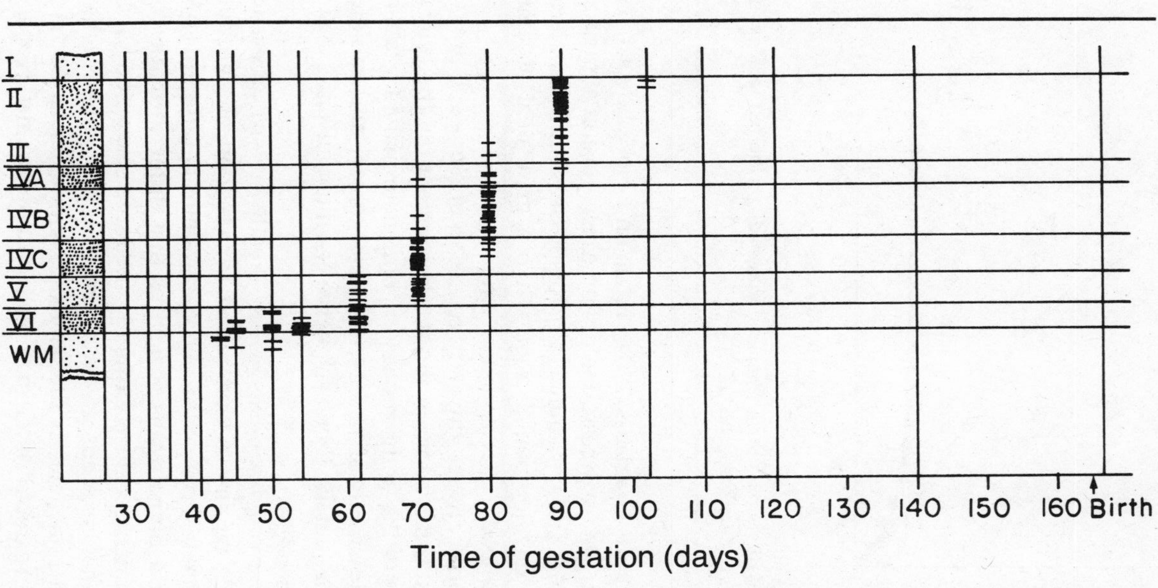

Figure 2.3. Generation of cortical neurons during the gestation of a rhesus monkey (which spans about 165 days). The final cell divisions of the neuronal precursors, determined by the incorporation of radioactive thymidine administered to the pregnant mother, occur primarily during the first half of pregnancy and are complete by about embryonic day 105. Each short horizontal line represents the position of the neurons labeled by maternal injection at the time indicated by the corresponding vertical line. The numerals on the left designate the cortical layers. (After Rakic, 1974)

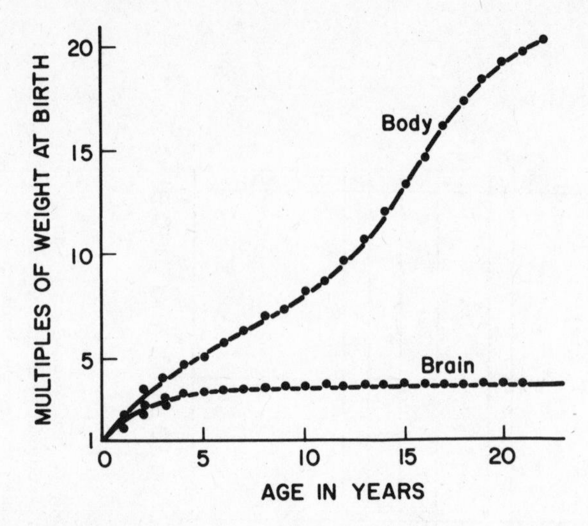

Figure 2.4. Growth of the human brain compared to overall growth of the body, using weight as a measure. (After Thompson, 1917)

Graziadei and Monti Graziadei, 1978; Rakic and Nowakowski, 1981; Bayer et al., 1982), an early cessation of nerve cell division in mammals is certainly the rule. Thus, mammals already possess a full complement of neurons at a developmental stage when a great deal of somatic growth still lies ahead.

This discrepancy between the early cessation of nerve cell proliferation and the continued addition to (or enlargement of) the cells that make up the rest of the body implies ongoing neuronal adaptation. If the original endowment of nerve cells is to control and monitor a growing body adequately, then each nerve cell must, in some manner, change commensurately.

The fact that related animals differ greatly in size presents a neurological issue similar to that raised by variation of size during development. That is, how do the nervous systems of different species manage to control homologous structures that differ so markedly in bulk? Most mammalian cells are roughly the same size among species of different dimensions. Thus the cells of most homologous organs in a shrew and a whale have nearly the same measurements, as do the cells of homologous organs in other mammals of different size (Table 2.1; Levi, 1905; Teissier, 1939; Altman and Dittmer, 1961). (As in development, nerve cells and muscle cells are again exceptional in that they *do* increase in size across species—see Chapter 4.) It follows, then, that larger mammals have greater bulk primarily because they harbor more cells. Thus, a 25-gram mouse comprises about 3×10^9 cells, whereas a 70-kilogram man comprises about 10^{13} cells (Baserga, 1985). Even among smaller invertebrates, many of which have quite small somatic cells compared to those of mammals and other vertebrates, constancy of cell size in the face of very different body sizes is apparent when related animals are examined (Conklin, 1911).

In the light of this information, one might expect the nervous systems of mammals or other vertebrates, by the measure of weight or cell number, to be proportional to body size. However, as noted early in this century, the size of the nervous system fails to keep pace with the size of the body in larger species (Donaldson, 1895; Hardesty, 1902; Thompson, 1917). This conclusion has since been confirmed in some detail. Measurements in a large number of mammalian species show that there is a progressive disproportion between the size of the nervous system (using brain weight as an index) and the body weight of animals of increasing size (Figure 2.5; Count, 1947; Jerison, 1961, 1973; Cobb, 1965; Gould, 1975). For example, the ratio of brain to body weight for a shrew is about 1:20, whereas for one of the larger whales this ratio is about 1:10,000. This failure of the size of the nervous

Table 2.1. Measurements of a variety of cell types indicating that there is little or no systematic variation in the size of most cell types among mammals of very different dimensions. Values are given as cross-sectional areas (μm^2), except in the case of red blood cells, where values are given as diameters (μm). (From Teissier, 1939; Altman and Dittmer, 1961)

Species	Liver cells	Thyroid epithelial cells	Renal epithelial cells	Pancreatic acinar cells	Red blood cells
Shrew	—	—	—	—	7.5
Mouse	390	123	177	211	6.6
Guinea pig	373	142	243	210	—
Rabbit	441	—	272	156	—
Cat	343	91	228	230	—
Dog	201	55	—	155	7.1
Pig	296	74	208	127	5.9
Ox	302	—	139	182	—
Horse	—	—	—	—	5.5
Elephant	—	—	—	—	9.2
Whale	—	—	—	—	8.2

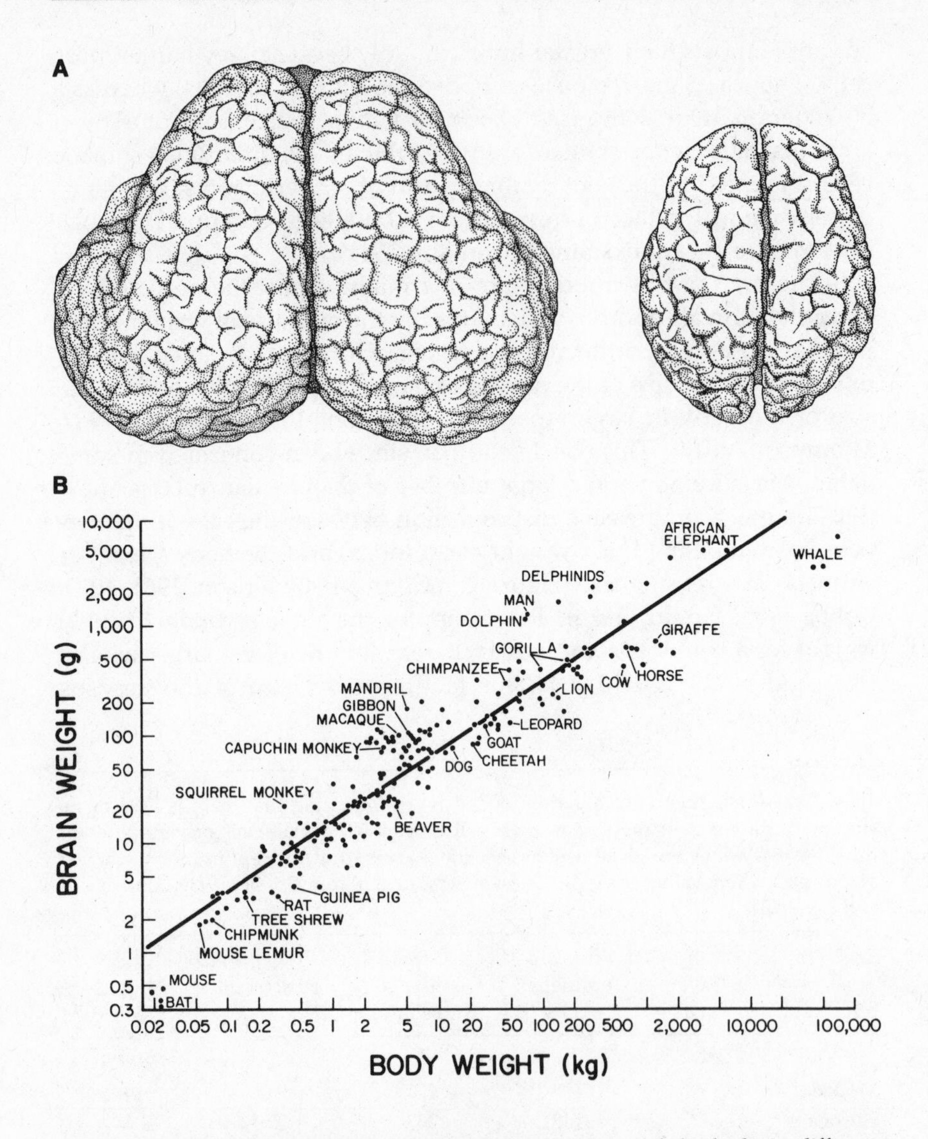

Figure 2.5. Size of the mammalian brain compared to size of the body in different species. *(A)* The dorsal aspect of the brain of an adult fin whale (left) and an adult man (right). The scale of the pictures is the same. Although the weight of a fin whale *(Balaenoptera physalus,* one of the largest cetaceans) is approximately 5,000 times that of a man, the brain weights differ by a factor of only about 5. *(B)* The relationship between the size of the body and the size of the nervous system, using brain weight and body weight as indexes for a large number of mammalian species. The straight line represents the solution of the "allometric equation" (Huxley, 1932). Whether the slope of this line is ⅔, which implicates body surface as a determinant of brain size, or ¾, which implicates metabolism as a determinant, remains unresolved (Jerison, 1961; Gould, 1975; Harvey and Bennett, 1983). (*A* after Tower, 1954; *B* after Count, 1947; Jerison, 1961)

system to keep pace with body size is all the more remarkable since the nervous systems of larger, more complex animals are responsible for a larger number of more sophisticated functions than are those of smaller species (primates, for instance, have a greater number of sensory processing areas in the brain than do smaller mammals; Kaas, 1977).

These comparisons of mammalian brain weight are complicated, to some degree, by the disproportionately large size of the cerebral cortex in man and other primates (presumably as a result of the evolution of "intelligence"). However, a similar discrepancy between somatic and neural size is apparent when populations of homologous nerve cells in the peripheral nervous system of different species are compared. An advantage of studies in the periphery is that the relatively small size of functionally similar neuronal populations allows a direct determination of the number of neurons in each species (in the brain, the numbers of nerve cells are too great to count directly and are assumed to change in proportion to brain weight). Quantitative comparisons across species can be made among the groups of spinal motor neurons that innervate particular skeletal muscles, among homologous sensory ganglia, and in the autonomic nervous system. Although only autonomic ganglia and the spinal neurons that innervate them have been studied systematically in this regard, scattered reports of spinal motor neuron pool size and of the numbers of neurons in the sensory ganglia of different animals are consistent with the findings in the autonomic system.

The number of cells in the superior cervical ganglion of mammals (Figures 1.1–1.2) ranges from about 10^4 in the mouse to about 10^6 in man, animals larger than man not having been examined (Table 2.2; Ebbesson, 1963, 1968a,b; Purves, Rubin, et al., 1986). As in the central nervous system, there is a progressive discrepancy between the number of nerve cells in these relatively simple relay stations and the overall size (by weight) of the animals in which the ganglia reside. For example, the body weights of a mouse and a rabbit differ by a factor of about 70; the numbers of superior cervical ganglion cells in these two species, however, differ by only a factor of 4. Body weights across the full range of animals studied—mouse to man—differ by a factor of about 2,400, whereas the numbers of nerve cells in the superior cervical ganglia increase by a factor of only about 90. The same sort of disproportion is evident among the spinal neurons that drive autonomic ganglion cells. Thus the number of nerve cells in the spinal cord of a rabbit that innervate the superior cervical ganglion is only about twice the number in a mouse (about 1500 neurons in a rabbit, compared to 700 neurons in a mouse), and the number of preganglionic neurons in a man is only about 5 times the number in a rabbit. Clearly, then, the numbers of mammalian autonomic nerve cells in both ganglia and spinal cord, like the numbers of neurons in the brain, do not keep pace with increasing body size among related species.

The relative shortfall of neurons in progressively larger species can be likened to the discrepancy between neuronal number and body size during the growth of a particular animal. In both instances, a relatively fixed number of nerve cells must monitor and motivate target structures that vary greatly in size. Even in those regions of the nervous system that are remote from the primary motor and sensory pathways, this constraint must, in some measure, be felt. The implication of these discrepancies is that nerve cells, unlike most other cell types, must be capable of ongoing adjustment to a changing body.

Table 2.2. Numbers of superior cervical ganglion cells and related preganglionic spinal neurons in mammals of different size. Counts of homologous neuronal populations in the peripheral nervous system indicate a disproportion similar to that observed between the number of central neurons (inferred from brain weight) and body size. Values for both weight and neuronal numbers are only approximate.

Species	Weight of animal (g)	Number of superior cervical neurons	Number of preganglionic neurons innervating superior cervical ganglion	Reference
Mouse	25	10,000	700	Purves, Rubin, et al., 1986
Hamster	100	17,000	700	Purves, Rubin, et al., 1986
Rat	200	26,000	1,000	Purves, Rubin, et al., 1986
Guinea pig	400	36,000	1,300	Purves, Rubin, et al., 1986
Rabbit	1,600	41,000	1,500	Purves, Rubin, et al., 1986
Cat	3,000	114,000	—	Levi-Montalcini and Booker, 1960
Macaque	4,000	243,000	4,900	Ebbesson, 1968a,b
Baboon	13,500	397,000	6,500	Ebbesson, 1968a,b
Chimpanzee	38,000	753,000	7,700	Ebbesson, 1968a,b
Man	70,000	911,000	8,300	Ebbesson, 1963

Influence of Form on the Nervous System

Variation in animal size, however, is only a part of the story. As Haldane pointed out, a change in size is usually accompanied by a change in form. The changing form of animals (which of course has many causes in addition to a change of size) places further demands on the plasticity of neurons and their connections.

To take an extreme case, a giraffe often grows 100 centimeters or more in height during the first six months of postnatal life, at weekly rates that may be as great as 20 centimeters (Dagg and Foster, 1976). The major part of this growth, not surprisingly, occurs in the animal's neck and legs. Accordingly, the innervation of targets must not only change to keep pace with this enormous growth but also make larger adjustments in the cervical and limb regions than at the level of the thorax.

Another dramatic example of altered form that has been studied in detail by neurobiologists concerns neuromuscular changes during metamorphosis. Metamorphosis, the transformation of a larva into an adult, often involves extensive changes in body form (as well as in organ physiology) in preparation for a new mode of existence. With respect to the innervation of muscle, a remarkable instance of such change occurs in the tadpole jaw, where the larval muscles degenerate and are replaced by an entirely new set with a different function and arrangement (Figure 2.6; DeJongh, 1968; Alley and Cameron, 1983). The jaw muscles, both larval and adult, are innervated by neurons in the trigeminal nucleus of the animal's brainstem. One might imagine that the nervous system would deal with such a radical change of form by simply eradicating the larval motor neurons and generating a novel set for the quite different task of operating the jaw of the adult frog (the larval jaw is specialized for respiratory aquatic feeding, whereas the adult jaw is specialized for catching insects). However, this is not what happens. The same motor neurons reorganize their peripheral (and presumably central) connections to deal with the novel requirements of the adult body (Alley and Barnes, 1983; Barnes and Alley, 1983).

The sensory system also provides examples of how the changing form of a developing animal requires the nervous system to change in tandem. Consider the visual system of certain flatfishes which spend their life on the ocean bottom and swim in a horizontal position (flounder, sole, and plaice, among others). In the larvae, the eyes of these animals are symmetrically placed on each side of the head, as in fish that swim in a vertical position. As these flatfish mature, however, one

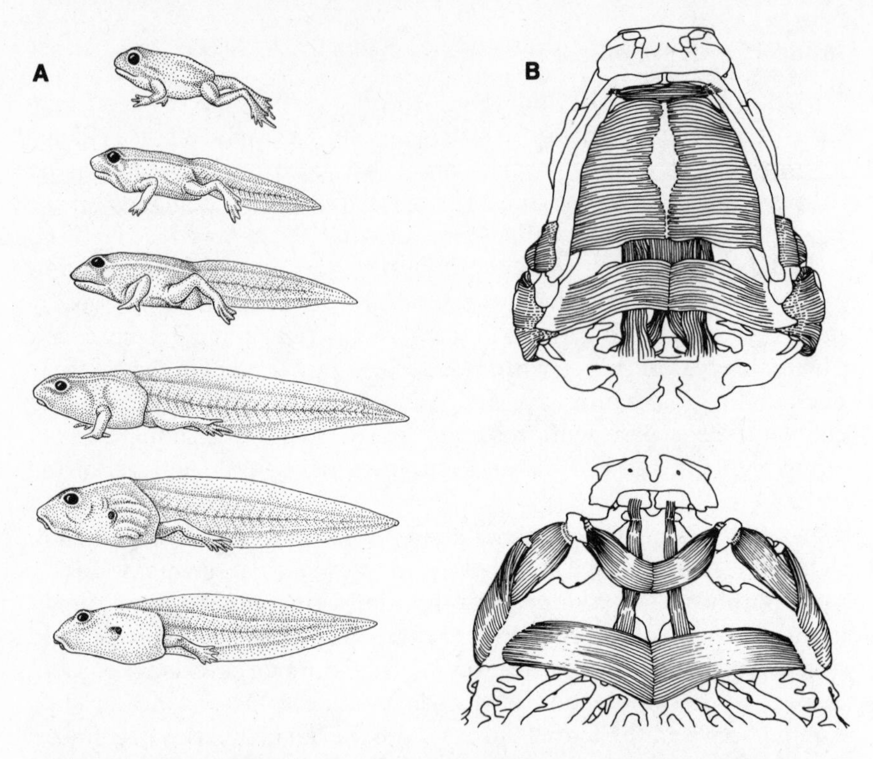

Figure 2.6. Metamorphosis in the frog *(Rana temporaria)* and some of its consequences for the neuromuscular system. *(A)* Stages of metamorphosis, from an immature tadpole (bottom) to a young frog (top). *(B)* Reconstruction of the jaw muscles in an immature tadpole (bottom) compared to the jaw and musculature of a young frog (top). The adult muscles differ from the tadpole muscles in structure and function but are innervated by the same neurons. (After DeJongh, 1968)

eye migrates so that both eyes ultimately lie on the side of the fish that is uppermost, which can be either the right or the left side, depending on the species (Williams, 1902; Kyle, 1921). As a result, the dorsoventral axes of the visual fields are ultimately perpendicular to the body, rather than being parallel to it, as in other teleosts (Luckenbill-Edds and Sharma, 1977). These changes in the position of the eyes with respect to the body require compensatory changes in the central connectivity of the visual system, as well as in related pathways such as the vestibulo-ocular system (Platt, 1973a,b; Luckenbill-Edds and Sharma, 1977; Graf and Baker, 1985a,b).

Another instructive instance of the demands of changing form on a sensory pathway occurs in the auditory system of the barn owl (Knudsen et al., 1977; Knudsen and Konishi, 1982). The owl uses binaural differences in the timing and intensity of sound to localize its prey in darkness (Knudsen et al., 1977), a feat that demands considerable precision. The animal's ability to make fine discriminations about the position of a sound source arises from the displacement of the two ears (Figure 2.7A). Because the ears are separated across the width of the face, auditory stimuli arrive at a slightly different time and with a slightly different intensity at each ear. The ears are also slightly displaced in the vertical plane, and the sound-collecting surfaces are tilted somewhat differently on the two sides (these surfaces, called the facial ruff, give the owl its characteristic appearance). Together, these features enable the owl to localize sounds in the vertical as well as the horizontal plane. As the owl matures, the anatomy of the face changes not only continuously but also disproportionately. That is, the facial ruff, the skull, and the separation of the ear canals all change at different rates (Figure 7B). These observations imply that the neural circuitry in the owl's auditory system continually adjusts to the changing geometry of the face to enable accurate sound localization. The same argument can be made for the visual system. In animals that have depth perception, for instance, the accurate interpretation of the retinal disparity that underlies stereopsis also depends on accurate neural compensation for the changing geometry of the face (Hubel et al., 1970; Bishop, 1987).

A similar phenomenon is apparent in phylogeny. That the nervous system must change in the course of evolution, in conjunction with the adaptive specialization of the rest of the body, is perhaps self-evident. Straightforward documentation of this idea is found in the overall structure of the nervous systems of highly specialized mammals. In general, the amount of the nervous tissue devoted to a given task is proportional to the sensory and motor apparatus given over to that function. Thus, among echolocating bats, a large fraction of the cerebral cortex is assigned to the analysis of the biosonar information which the bats use to locate prey and avoid obstacles in the dark (Polyak, 1926; Grinnell, 1963; Griffin, 1974; Suga, 1984). In rats and mice, a substantial part of the somatosensory cortex is allotted to an analysis of information from the prominent facial whiskers that these animals use to feel their way through narrow passages (Woolsey and Van der Loos, 1970). Among carnivorous mammals, those parts of the

A

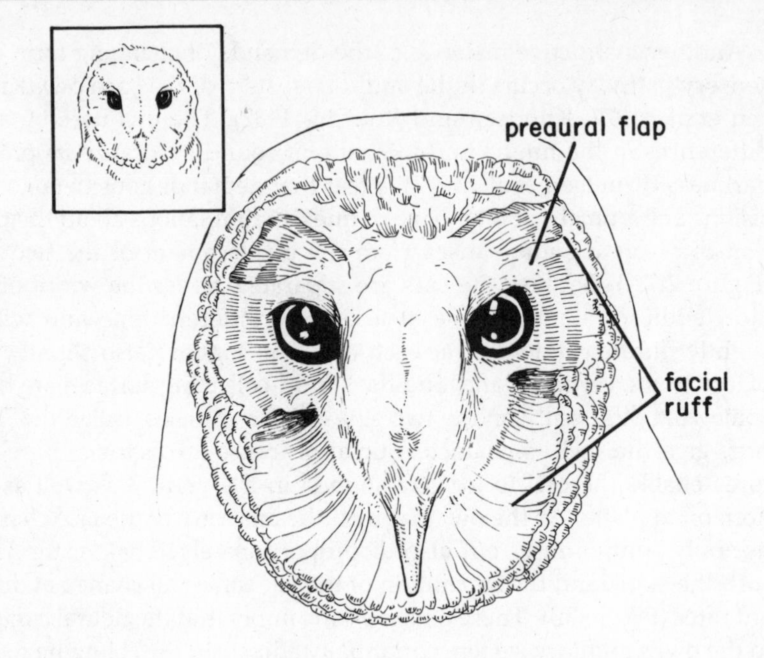
preaural flap
facial ruff

B

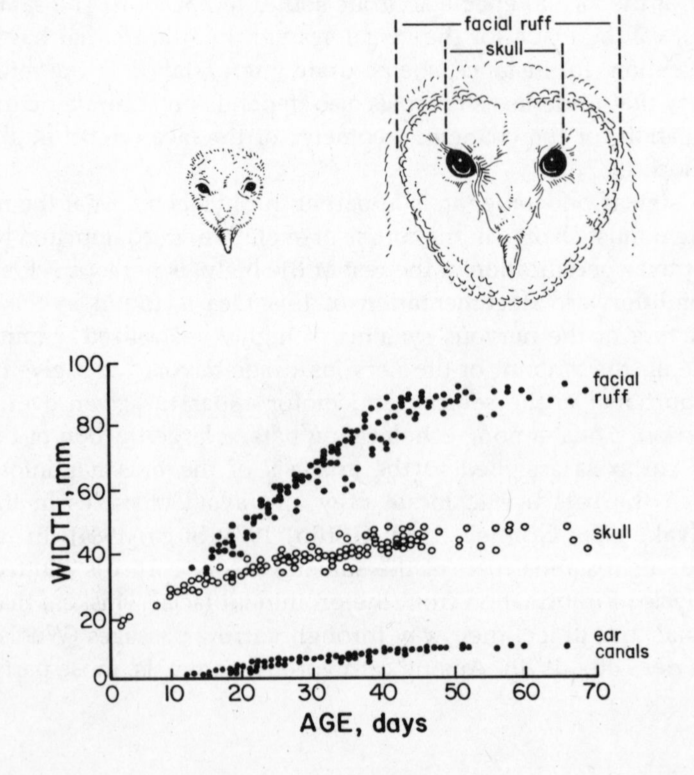
facial ruff
skull
WIDTH, mm
AGE, days
facial ruff
skull
ear canals
0
20
40
60
80
100
0
10
20
30
40
50
60
70

brain that have to do with olfaction are well developed, whereas in aquatic mammals that have little use for a sense of smell the homologous structures are usually rudimentary (Jacobs et al., 1979). Perhaps the most striking example of neural adaptation to a specialized somatic feature occurs in the teleost fish *Heterotis niloticus*. This animal has an elaborate paired spiral pharynx, which is richly innervated by the vagus nerve (Braford, 1986). Corresponding to this peripheral specialization is a marked enlargement of the vagal lobes of the animal's brain, which are also in the form of spirals.

We tend to think of the mechanisms that coordinate these parallel changes in the body and the related parts of the nervous system as opaque processes, obscured by the immense time scale of evolution and the persistent ignorance about the relation of genotype and phenotype in vertebrates. However, every animal breeder knows that substantial changes in animal size and form can be effected in just a few generations by careful selection and crossing of individuals that show a desired trait. In at least one instance, neurobiologists have made use of this fact to explore the neural correlates of phenotypic variation from one generation to the next. In outbred mice, the conventional arrangement of the facial whiskers referred to earlier is five rows on each side of the snout, with a defined number of whiskers (5 to 9) in each row (Figure 2.8). Beginning with an outbred stock, about a dozen strains of mice that had an abnormal whisker pattern were generated over just a few years; some strains had a smaller number of whiskers than normal, and others had a larger number (Van der Loos and Welker, 1985; Van der Loos et al., 1986). In mice and rats, the whiskers are represented within the brain by a special region of the somatosensory cortex called the barrel-field (Woolsey and Van der Loos, 1970), a name derived from the fact that a barrel-shaped agglomeration of cortical neurons is devoted to analyzing the sensory input from each whisker. It thus was

Figure 2.7. Changing arrangement of the ears and sound-collecting surfaces of the developing barn owl. *(A)* The superficial feathers are removed to show the external ear and the preaural flaps which cover the openings to the ear canals (the inset shows the owl's normal appearance). The ear opening on the left is slightly higher than the one on the right, a feature that helps the owl localize sound in the vertical plane. Sound localization is possible because differences in the time of arrival and intensity of a sound at the two ears vary as a function of both the lateral position and the height of the source. *(B)* The geometry of the face changes differentially as the animal grows. If accurate localization of sound is to occur during this period, the related neural circuits must adjust to the changing shape of the face. (After Knudsen, 1980; Knudsen, Esterly, and Knudsen, 1984)

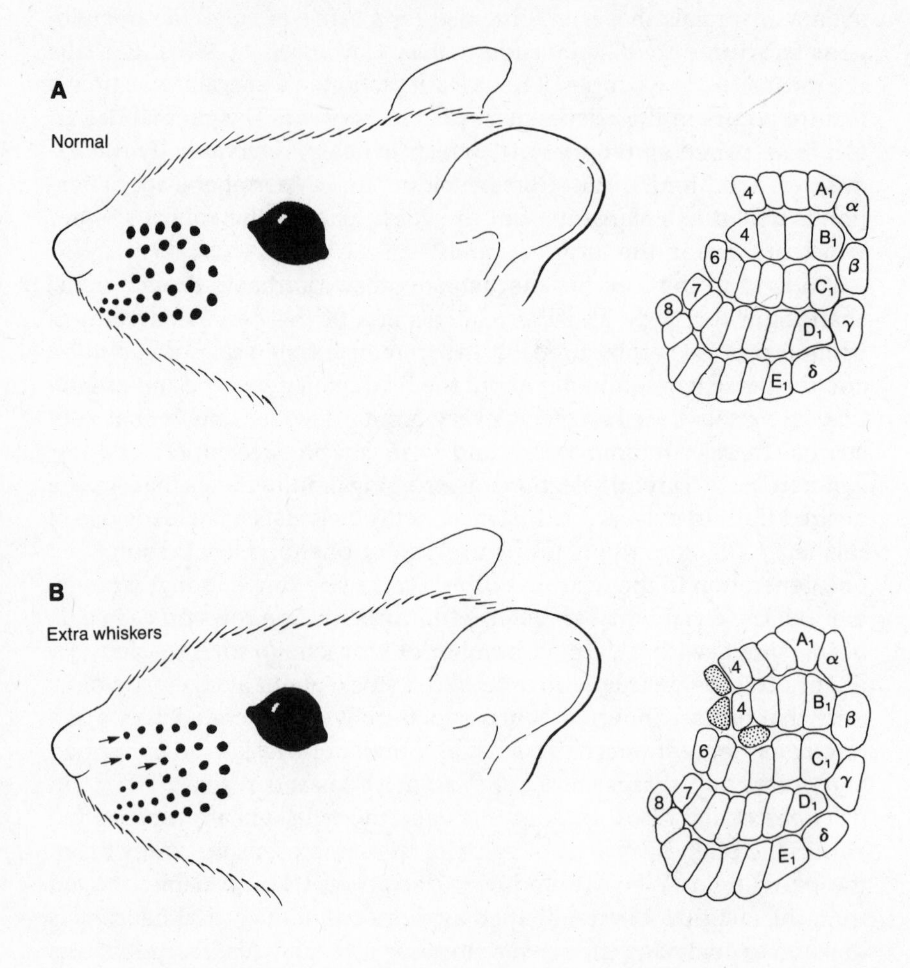

Figure 2.8. Representation of facial whiskers in the central nervous system of outbred mice and mice selectively bred for supernumerary whiskers. *(A)* Normal mice have five rows of whiskers, with a defined number of whiskers in each row (indicated by dots). In the corresponding somatosensory cortex (right), each whisker is represented by a cluster of neurons called a barrel (the outlines of the barrels are drawn from histological sections of the brain after routine staining). The entire group of barrels is called the barrel-field. The numbers and letters represent the conventional nomenclature. *(B)* Mice can be selectively bred for a greater or smaller number of whiskers than normal. This strain of mouse has three extra whiskers on each side of the snout—one occurring in the first and second rows, another lying between the second and third rows (arrows). In such strains, the cortical representation of the whisker pad includes a supernumerary cortical barrel (stipple) for each extra whisker. (After Van der Loos et al., 1984)

possible to ask whether strains of mice with more, or fewer, whiskers reflect this somatic variation in the organization of the barrel-field. The answer was clear: mice with extra whiskers have extra barrels, and those with fewer whiskers have correspondingly fewer barrels (Van der Loos and Dörfl, 1978; Yamakado and Yohro, 1979; Van der Loos et al., 1984; Welker and Van der Loos, 1986).

These findings, together with the general observation that the organization of the brain in various specialized animals reflects the organization of the body, imply a means of communication between the body and the nervous system which is capable of modulating neural organization from generation to generation. The alternative possibility of preprogrammed, parallel changes in the nervous system and its targets seems improbable on the face of it and is belied by a wide range of developmental studies which indicate a marked dependence of the neural structures on their targets (see Chapters 3–5). If, for example, a whisker is removed from the face of a mouse early in development, then the corresponding barrel in the somatosensory cortex is also lost, even though there are several synaptic links between the neurons in the brain and the whisker itself (Van der Loos and Woolsey, 1973; Belford and Killackey, 1980; Jeanmonod et al., 1981; Woolsey et al., 1981). Although it is equally unlikely that the mouse cortex is simply a *tabula rasa* awaiting instructions from the periphery, there is no doubt that the nervous system can adjust to somatic variation in a highly coordinated manner from one generation to the next.

Significance of Neural Adjustment to Changes in Size and Form

These various examples indicate that differences in the size and form of animals, whether in ontogeny or in phylogeny, are accompanied by commensurate changes in the nervous system that ensure efficient neural control of the body. Although the examples do not indicate *how* neural adjustments to the requirements of the body might occur, they make two general points which are relevant to the mechanisms involved. First, these observations indicate the existence of regulatory mechanisms operating between neurons and the cells that they innervate (whether within the nervous system or outside it). These interactive mechanisms appear to perform equally well in the service of ontogeny or phylogeny. In both instances, their operation allows the nervous system to adjust to a changing body with a minimum of genetic instructions.

A second general point is that neural adjustment to changing size and form involves coordinated changes in the *connections* of nerve cells, in addition to changes in the number of available nerve cells. In neither development nor speciation do the numbers of nerve cells available change in proportion to the body they must innervate. Why Nature has not chosen simply to add or subtract a proportionate number of neurons in animals of different size and form is not known. However, one implication of this strategy is apparent: nerve cells must adjust to changing bodies by altering the number of their axonal and dendritic branches and the synapses that these processes make.

THREE

Coordination of Neuronal Number and Target Size

DIFFERENCES in size and form of vertebrates during maturation, and among related species, indicate that the nervous system must adjust to the needs of a changing body. Although there are many ways that such neural malleability might be realized, changes in two parameters are fundamental: variation in the number of neurons available (which does change among animals, albeit disproportionately) and variation in axonal and dendritic branching (and therefore synaptic connectivity). This chapter is concerned with the first of these issues, the establishment of appropriate numbers of nerve cells available for various functions. Historically, it is this aspect of neural development that has established the importance of trophic interactions in vertebrates. The following chapters extend the concept of trophic interaction as a basis for neural malleability from the regulation of neuronal numbers to the regulation of neuronal branches and the synaptic connections they form.

Neuronal Populations in Small, Simple Animals

In the smallest animals that have nervous systems, the phrase *neuronal populations* does not have the same meaning that it has in large animals. In some minuscule invertebrates—the 0.5-millimeter-long roundworm *Caenorhabditis elegans* is the most thoroughly studied example (Figure 3.1A)—each animal comprises a precise number of cells. Of 959 somatic cells in the mature hermaphrodite, 302 are nerve cells (Sulston and Horvitz, 1977; Horvitz, 1981). In such tiny invertebrates, each nerve cell has a well-defined function and is, in this sense, unique.

For larger, more complicated animals the situation is different. In mammals, the number of somatic cells is vastly larger—on the order of 10^9 for a mouse and 10^{13} for a man. In such animals, each muscle is served by tens or hundreds of primary motor neurons, and each region

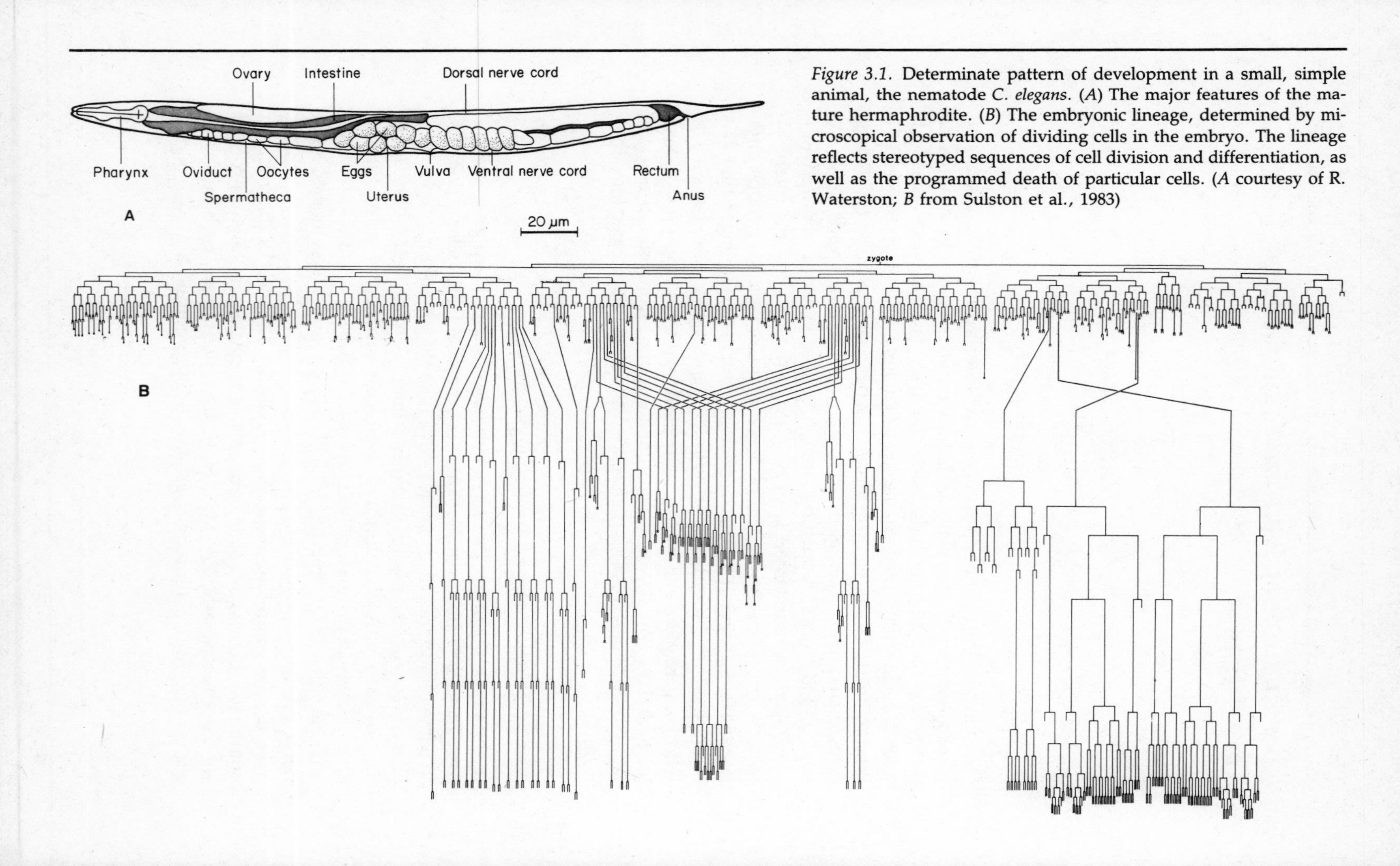

Figure 3.1. Determinate pattern of development in a small, simple animal, the nematode *C. elegans*. (*A*) The major features of the mature hermaphrodite. (*B*) The embryonic lineage, determined by microscopical observation of dividing cells in the embryo. The lineage reflects stereotyped sequences of cell division and differentiation, as well as the programmed death of particular cells. (*A* courtesy of R. Waterston; *B* from Sulston et al., 1983)

of the body is innervated by many primary sensory nerve cells. Nerve cells not belonging to the primary motor and sensory divisions of the nervous system (interneurons and local circuit neurons) also increase from small numbers in the simpler invertebrates to extremely large populations in the central nervous system of mammals. Thus, in progressively larger and more complex animals, increasingly large ensembles of neurons perform the functions served by one or a few nerve cells in very simple animals. Moreover, the number of nerve cells given over to particular functions in larger animals varies substantially among individuals of the same species.

The problem of establishing the right number of nerve cells to deal with various functions admits, at least in principle, a relatively straightforward solution: a developmental program might simply generate a predetermined number of neural and non-neural cells, each with a highly specific role. In fact, a determinate strategy of this sort is used by some small, relatively simple animals. In the case of the nematode *C. elegans*, the entire sequence of development has been painstakingly catalogued by microscopic observation of the division and differentiation of cells in the translucent embryo (Brenner, 1973, 1974; Sulston and Horvitz, 1977; Kimble and Hirsh, 1979; Horvitz, 1981; Sulston et al., 1983; Chalfie, 1984). This work has provided the first example of an animal in which the fate of every cell from fertilization to maturity, including the nerve cells, is known (Figure 3.1B). The analysis of nematode development has led to some powerful conclusions. Foremost among these is the finding that the ancestral history of each cell in such animals is normally invariant. This fact is all the more remarkable because of the death of some of the cells in the embryonic lineage. Thus the generation of the 959 adult cells is accompanied by the predictable death of 131 other cells (Sulston and Horvitz, 1977; Sulston et al., 1983; Ellis and Horvitz, 1986). A second conclusion is that lineage is a major determinant of both the function and the position of cells. What a cell becomes, and where it ultimately resides, are the result of its line of descent.

Cellular pedigrees, however, are not immutable during the maturation of even the simplest animals. If, for example, the development of *C. elegans* is perturbed by ablating a particular cell (usually done with a laser microbeam), the fate of other members of that lineage, or the lineage of neighboring cells, may be altered (Kimble et al., 1979; Sulston and White, 1980; see also Blair, 1983; Taghert et al., 1984; Doe et al., 1985; Sternberg and Horvitz, 1986). Similarly, the elision of a portion of the pedigree by genetic means in certain mutant strains of C.

elegans results in altered fates for some cells (Sulston and Horvitz, 1981; Fixsen et al., 1985). Thus, the development of even this extremely small and simple animal admits some variability, as well as the ability to respond to perturbations by means of cellular interactions.

Nor is the largely determinate development of *C. elegans* typical of invertebrates. Embryogenesis in nematodes and insects, for instance, is different in fundamental ways (Wieschaus and Gehring, 1976; Stent and Weisblat, 1985). In nematodes (and in some segmented worms, such as leeches), determinate lineages can quite literally be traced back to the uncleaved egg. In insects, however, lineage appears to become important only after the blastula stage, when several thousand nominally equivalent nuclei migrate to the periphery of the embryo and adopt more restricted fates (as well as cell membranes). Phyletic comparisons are further complicated by the fact that the term *invertebrate* alludes to a vast range of animals, including many large, structurally complex species. The cephalopod *Octopus vulgaris*, for instance, commonly weighs a kilogram or more, lives for several years, shows a wide range of behavior, and has a large "brain" (Young, 1963; Wells, 1978). In contrast to *C. elegans*, whose 302 nerve cells form about 7,000 synapses, the nervous system of *O. vulgaris* contains some 500 million nerve cells which make about 10^{11} synapses (Young, 1963). The giant Pacific octopus, *Octopus honkongensis*, whose size is typically 10–100 times greater than that of the common octopus presumably has a much larger and more complex nervous system. The developmental strategy of an animal like the octopus (which has never been explored) is probably much more like that of vertebrates than that of a fellow invertebrate such as *C. elegans*. Thus, distinctions based on size and complexity may be more germane here than distinctions based on the presence or absence of a backbone. Nevertheless, the elaboration of the adult complement of nerve (and other) cells in at least some very small invertebrates proceeds according to largely determinate programs in which feedback and adjustment is a relatively minor, though important, theme.

Neuronal Populations in Large, Complex Animals

In large, complex animals like mammals and other vertebrates, the establishment of appropriate numbers of nerve (or other) cells by a fixed lineage strategy (or any determinate program) is much less plausible. Whereas the size and form of individual representatives of a species like the nematode *C. elegans* are largely invariant, for mammals and other vertebrates, size and form—and therefore cell number—are highly variable among individuals. This fact implies that cellular interactions play a more prominent role in establishing cell populations in large, complex animals.

The present consensus that vertebrate ontogeny is to some degree indeterminate is based primarily on classical studies of the phenomena of induction and regulation (Hamburger, 1988). Induction, in which the fate of embryonic cells is influenced by signals arising from other nearby cells, and regulation, in which an embryo compensates for a missing portion after experimental ablation, imply a good deal of flexibility in vertebrate development in general and the nervous system in particular (Jacobson, 1978; Purves and Lichtman, 1985b). The inference that vertebrate development is in part indeterminate has been strengthened by studies which are similar, in principle, to those on the development of simple invertebrates. Much of this work has been carried out in the zebrafish (Figure 3.2A; Kimmel and Warga, 1986, 1987a,b). An advantage of this animal is that, as in the nematode, the early embryo is translucent; it is thus possible to follow the lineage of individual cells by direct observation. In the larger and more complex zebrafish embryo, however, the task of cell tracing is facilitated by the use of an intracellular marker (Weisblat et al., 1978). An individual cell (blastomere) of known identity is injected with a fluorescent dye shortly after the fertilized egg has cleaved (Figure 3.2B). The persistence of the nontoxic dye in the progeny of the injected cell enables lineage tracing among the more numerous cells of the fish embryo; in this way the fates of clonally related cells can be catalogued with respect to their number, position, and the tissue or organ to which they contribute.

Unlike the nematode, in which the fate of the progeny of particular cells is highly predictable, the progeny of the same blastomere in different fish embryos comprise cells that vary markedly in number, position, and functional type (Figure 3.2C; Kimmel and Warga, 1987a,b; see also Winklbauer and Hausen, 1985a,b). Evidently the construction of vertebrates proceeds in a more flexible way than the construction of some very small and simple invertebrates. However, just as the smallest invertebrates manifest a measure of indeterminacy in their cell lineages, so vertebrates give some evidence of determinacy. Thus similar lineage studies in the toad *Xenopus laevis* show that there is some tendency for the progeny of each blastomere to populate a defined region of the late stage embryo (Jacobson, 1980; Klein and Moody, 1987; Moody, 1987a,b; Sheard and Jacobson, 1987). Moreover, if cells in

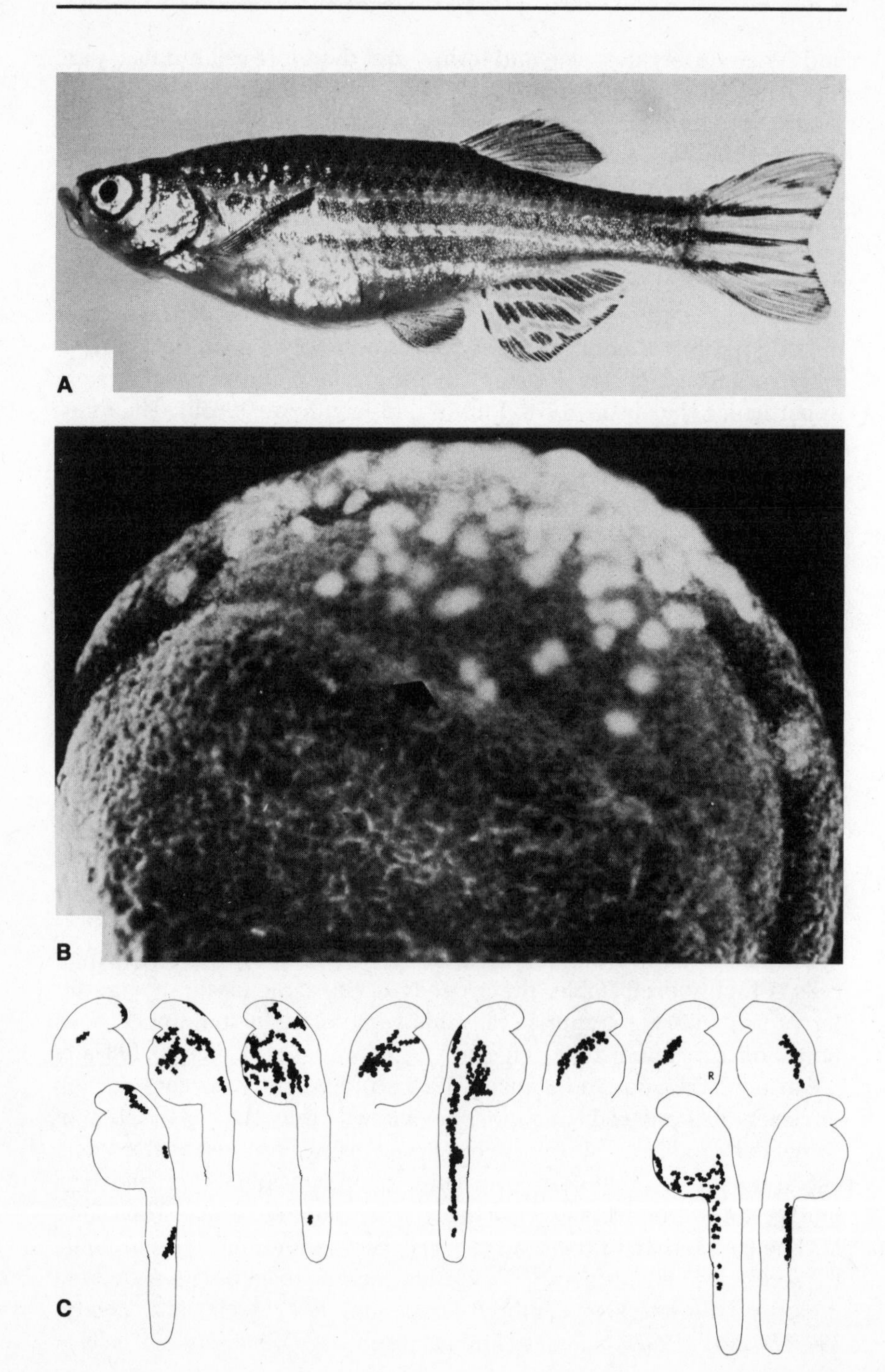
A
B
R
C

the zebrafish are labeled at a somewhat later stage of development, a greater restriction of cell fates is apparent (Kimmel and Warga, 1986).

That the development of mammals and other vertebrates is in some measure a flexible process, in which the outcome is not fully determined before the fact, is also apparent in instances where lineage studies from such early embryonic stages are impractical. Neurobiologists who use laboratory mammals (and neurologists and neurosurgeons who deal with patients) take for granted a certain amount of variation in the structure of the nervous system. For example, variations in the size, number of constituent neurons, and overall arrangement of autonomic ganglia among individuals of the same species are notorious (Pick, 1970; Gabella, 1976). In the system of spinal sensory ganglia, in which the number of neurons can also be counted directly, differences in the numbers of neurons among normal mice of the same strain range up to 30 percent (Levi and Sacerdote, 1934). A similar diversity of size and detailed arrangement is apparent in the central nervous system (Williams and Herrup, 1988). In the brainstem nuclei, where the numbers of nerve cells are still relatively small, counts of homologous populations show substantial differences from animal to animal of the same species (Herrup et al., 1984). Similar counts in the brain itself are much more difficult because of the large number of cells involved. However, variation is readily apparent. The average weight of the brain in an adult human (male) is about 1,400 grams (Pakkenberg and Voigt, 1964; Dekaban and Sadowsky, 1978). The range of "normal" human brain weight, however, is at least ± 200 grams and possibly much more (Donaldson, 1895; Le Gros Clark, 1959; Altman and Dittmer, 1962; Cobb, 1965; Dekaban and Sadowsky, 1978). Such variations presumably reflect substantial differences in the number (and size) of the cells that constitute the brain (and large differences in the number

Figure 3.2. Indeterminate pattern of development in the zebrafish (*Brachydanio rerio*). (*A*) An adult zebrafish. The overall length of the animal is about 3 centimeters. (*B*) A zebrafish gastrula, showing the progeny of a single blastomere injected at an earlier stage with a fluorescent dye. The dye is retained in the descendants of the injected cell, which can thus be followed as development proceeds. (*C*) Location of the descendants of the same labeled blastomere in ten different embryos at a still later stage. In the eleventh case, the embryo labeled "R," the injection was made into a blastomere that immediately neighbored the labeled cell in the other examples. All of the labeled cells are in the embryo's surface epithelium. The wide and variable dispersion of the clonally related cells differs sharply from the strict and largely invariant relationship among progeny observed in the development of the nematode *C. elegans*. (*A* courtesy of H. M. Howard; *B* from Kimmel and Warga, 1986; *C* from Kimmel and Warga, 1987b)

of neuronal connections—see Chapter 4). In spite of this situation, small-brained people are not thought to be less intellectual than individuals who have large brains. Since no modern study correlating brain size and function has been done, this point is usually made by citing the reported brain weights of various historical figures (Donaldson, 1895; Gould, 1981). The brain of Anatole France, for example, is said to have weighed only 1,017 grams. An extreme example is that of a 46-year-old watchman at New York's Penn Station who died suddenly of asphyxia; although he had allegedly performed his job satisfactorily until the time of his death, the medical examiner found his brain to weigh 680 grams (Wilder, 1911). In contrast, the brain of the naturalist Georges Cuvier weighed 1,830 grams, and that of the Russian novelist and poet Ivan Turgenev weighed a remarkable 2,012 grams. Although reports of this sort cannot be taken at face value (variables such as age, body weight, and general health are not taken into account), there is no doubt that the overall size of normal human brains varies greatly without obvious effect on mental performance.

Another approach to the assessment of normal variability in the human brain has been provided by physicians interested in delimiting, for clinical purposes, brain regions devoted to a particular function. When a portion of the brain is to be removed, the surgeon must meticulously delineate regions which, if damaged, would leave the patient with a significant deficit. One such region is that devoted to language, traditionally considered to be the area described by P. Broca and C. Wernicke in the lateral cortex of the left hemisphere for a right-handed person (Geschwind, 1970). However, careful electrical mapping of this region during surgery performed under local anesthesia shows a great deal of variation in the arrangement of the language area in different individuals (Ojemann, 1983). Clinical observations of this sort have been confirmed by more detailed experimental work on other primates. Careful mapping of the cortical areas that subserve a particular somatosensory modality in the monkey, for example, shows substantial differences in the arrangement of the same functions from animal to animal (Figure 3.3; Merzenich, 1985; Merzenich et al., 1987).

Each of these approaches to assessing normal variation in the vertebrate nervous system has serious limitations. Nevertheless, each study points to the same conclusion: the size and arrangement of the nervous system is not identical among different individuals of the same species. This conclusion is the neurological equivalent of a fact that is taken for granted with respect to other, more readily observable features, namely variation among individuals. Whether within the nervous sys-

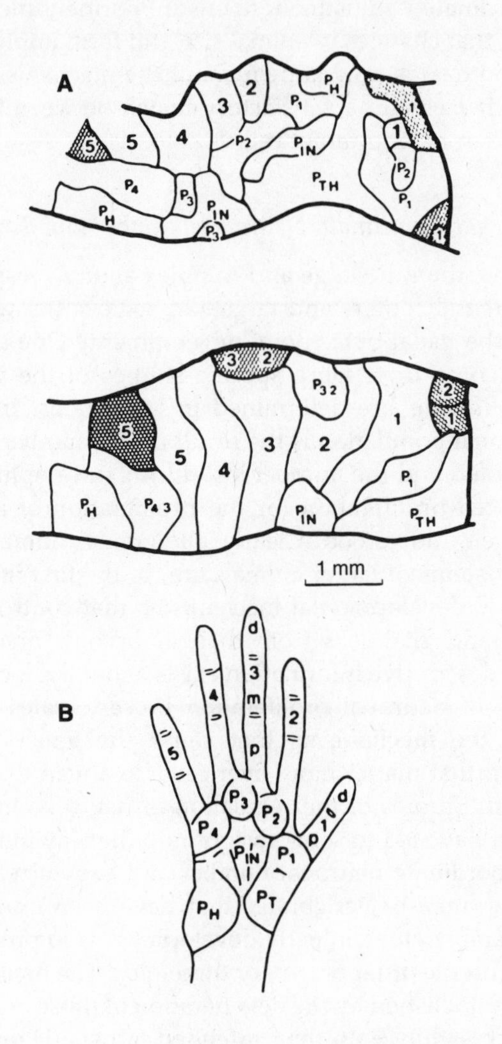

Figure 3.3. Regions of the somatosensory cortex that responded to tactile stimulation of the hand in two different squirrel monkeys. (*A*) Maps made by recording the location of electrical responses in the cortex evoked by stimulating various parts of the contralateral hand. (*B*) Regions on the palmar surface that elicited the responses indicated in the cortical maps (the stippled sectors in *A* represent responses to touching the dorsal surface of the hand). Variations of this magnitude among different individuals of the same species were typical of a large number of animals examined. (After Merzenich et al., 1987)

tem or outside of it, phenotypic variation in vertebrates implies a means of coordinated adjustment to ensure normal function (in much the same way that changes in animal size and form imply such mechanisms). A good deal is now known about the mechanisms in the nervous system that ensure a satisfactory match between indeterminate numbers of neurons and target cells.

Mechanisms That Coordinate Neuronal Number and Target Size

By what means, then, do large and complex animals respond to variation in both neural centers and targets to achieve the well-integrated result that is the usual outcome of development? One answer to this question has come from more specific studies of the way in which neuronal populations are determined in vertebrates. In general, the size of a neuronal population is the result of the number of nerve cells initially generated and the number lost during development. The control of nerve cell proliferation (or the proliferation of any other cell type) is not well understood. Clearly, the greater numbers of cells in the nervous systems of larger animals are, in the first instance, determined by those developmental mechanisms that control the elaboration of nerve and glial cells from their embryonic precursors. Even though much descriptive information exists about the location, timing, and sequence of neuronal proliferation in vertebrates (Sidman and Rakic, 1982), the mechanisms that cause the generation of more neurons (or for that matter more liver cells) in a man than in a mouse are not known. Studies of the regulation of neuronal number by cell loss, however, have led to some insight into the way in which cellular interactions coordinate neuronal number and body size.

In the wide range of vertebrates that have been examined in this regard, a general strategy in early development is to produce a surfeit of nerve cells (on the order of two- or threefold). The final population is subsequently established by the degeneration of those neurons that fail to interact successfully with their intended targets (Hamburger, 1977; Oppenheim, 1981; Cunningham, 1982; Hamburger and Oppenheim, 1982; Cowan et al., 1984; Oppenheim, 1985; Purves and Lichtman, 1985b). Although most detailed descriptions of this process concern those motor and sensory neurons that are directly associated with somatic targets, the strategy at this level also applies to neurons that are not in direct contact with muscles, glands, or sense organs.

Evidence that targets influence the size of neuronal populations in vertebrates comes from a long series of studies that began shortly after the turn of the century. The seminal observation was that the removal of a limb bud from a chick embryo results, at later embryonic stages, in a striking reduction of the numbers of nerve cells in the corresponding portions of the spinal cord (Figure 3.4; Shorey, 1909). This and related experiments carried out in birds and amphibians in the 1920s and 1930s confirmed the dependence of central neurons on the presence and extent of the periphery (Detwiler, 1920, 1936; Hamburger, 1934, 1939a,b).

For several decades, the degeneration of neurons following early target ablation was considered to show that the periphery somehow controlled neuronal proliferation. It was not until a reassessment of the matter in the 1940s that the degeneration of some nerve cells was recognized as a *normal* developmental phenomenon (Levi-Montalcini and Levi, 1942, 1944; Hamburger and Levi-Montalcini, 1949, 1950). In the motor system, for example, an excess of skeletal and visceral motor neurons is initially generated in the spinal cord. As central neurons associate with limbs or autonomic targets that are distributed non-uniformly along the rostrocaudal axis, corresponding differences among the original population appear in the spinal cord. Thus, larger numbers of motor neurons survive in the brachial and lumbar regions of the cord because the limb innervation derives from these levels. Similarly, preganglionic neurons that innervate sympathetic ganglia survive at the levels where the target ganglia are located, but they succumb at other levels (Hamburger and Levi-Montalcini, 1950). Subsequently, experiments in a variety of neural regions and species have shown that the normal death of part of the original population of neurons is widespread, if not ubiquitous, in the development of vertebrates (Hamburger, 1975; Oppenheim, 1981; Cunningham, 1982; Hamburger and Oppenheim, 1982; Cowan et al., 1984; Oppenheim, 1985; Purves and Lichtman, 1985b).

A number of explanations of neuronal death in vertebrates have been proposed, including the selective demise of neurons predestined to die through some intrinsic failing, the selective death of neurons that fail to extend an axon to their intended target, and the selective death of neurons that happen to reach the wrong destination. Although each of these explanations may apply to special cases, none is adequate as a general rule. First, neurons in developing populations appear to differentiate and mature normally until the period of cell death (Landmesser and Pilar, 1974; Pilar and Landmesser, 1976; Chu-

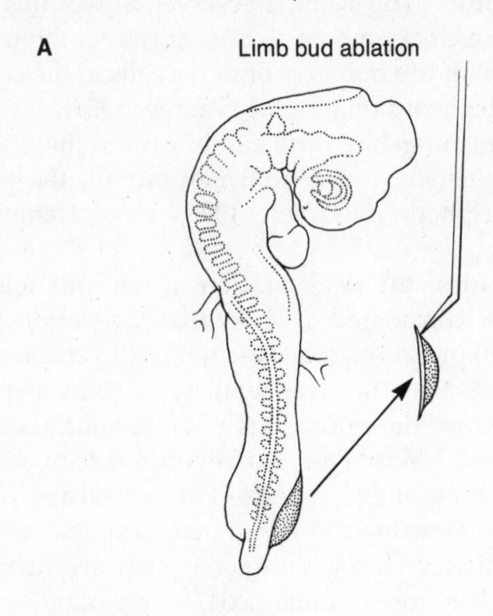

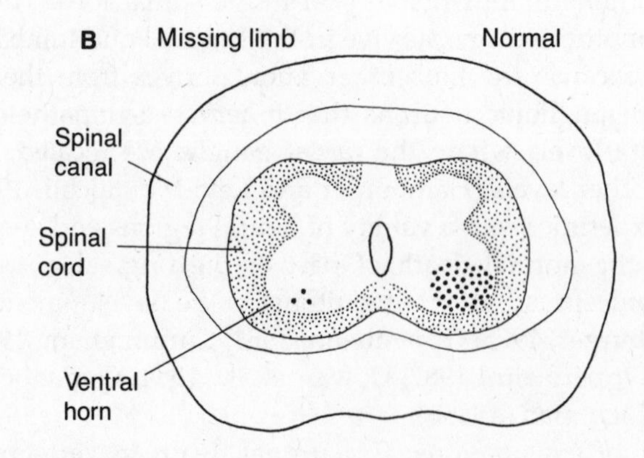

Figure 3.4. Effect of removing the target on the survival of related neurons. (*A*) Limb bud amputation in a chick embryo at the appropriate stage of development (about 2.5 days of incubation), which results in a marked depletion of the motor neurons that would have innervated the missing extremity. (*B*) Cross-section of the lumbar spinal cord of an embryo that underwent this surgery about a week earlier. The motor neurons (dots) in the ventral horn that would have innervated the hindlimb degenerate almost completely after embryonic amputation; the normal complement of motor neurons appears on the other side. (After Hamburger, 1958, 1977)

Wang and Oppenheim, 1978; Oppenheim et al., 1978). It seems unlikely, therefore, that cell death represents the demise of an intrinsically unhealthy subgroup. Second, a marker substance like horseradish peroxidase (a protein which neuronal processes can take up and transport back to their cell somata) retrogradely labels all the neurons in the relevant population when introduced at the level of the target before the period of cell death (Cowan and Clarke, 1976; Chu-Wang and Oppenheim, 1978; Pilar et al., 1980). This result indicates that the neurons which die are not ones that simply failed to reach their target. The timing of neuronal death is consistent with this idea. The death of motor neurons in the chick, for example, occurs during the fifth to the tenth day of the 21-day period it takes a chick to develop and hatch; in a mammal like the rat, the phase of neuronal degeneration is also limited to midembryonic life (Chu-Wang and Oppenheim, 1978; Hardman and Brown, 1985). The timing of neuronal death in these animals corresponds roughly to the period during which axons grow out from nerve cells and make contact with their normal targets. In instances in which the period of axonal outgrowth is drawn out—for example, in the sympathetic ganglia of mammals (Hendry and Campbell, 1976; Rubin, 1985a,b)—the phase of neuronal degeneration is also prolonged (Nornes and Das, 1974; Hendry, 1977). Finally, neurons do not die because they have reached the wrong destination. Although in some instances in the central nervous system death follows erroneous projection (O'Leary et al., 1986; O'Leary, 1987), in other systems, growing axons project accurately to appropriate targets before the period of cell death (Landmesser, 1980; Westerfield et al., 1986).

The explanation for the majority of neuronal death in early vertebrate development appears to be that neurons fail to interact successfully with their targets. More specifically, they fail in competition with their peers for support derived from the neural target. Competition, in this context, is defined in much the same way as in ecology, namely as the "active demand by two or more organisms [here nerve cells] for a common resource" (Wilson, 1980, p. 42). The resource that neurons seek to acquire is thought to be a class of molecular signals called trophic factors (Chapter 7).

Evidence for the competitive nature of the process that leads to the death of a portion of each initial population of neurons comes from experiments in which the ratio of neurons to target cells is decreased. Experiments of this sort show that many neurons that would normally have died can be salvaged by augmenting the amount of target available (Figure 3.5). One way to do this is to implant a supernumerary

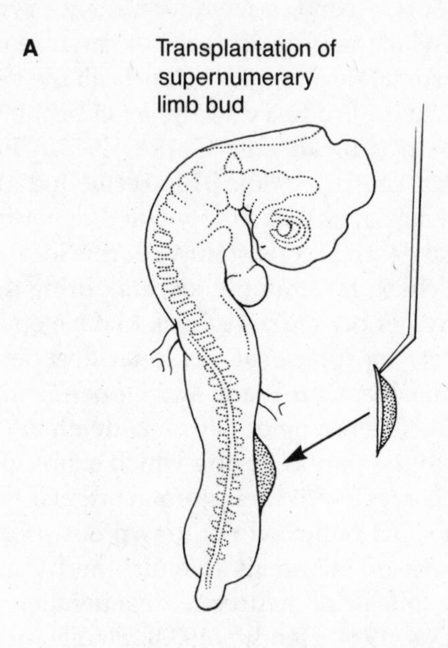

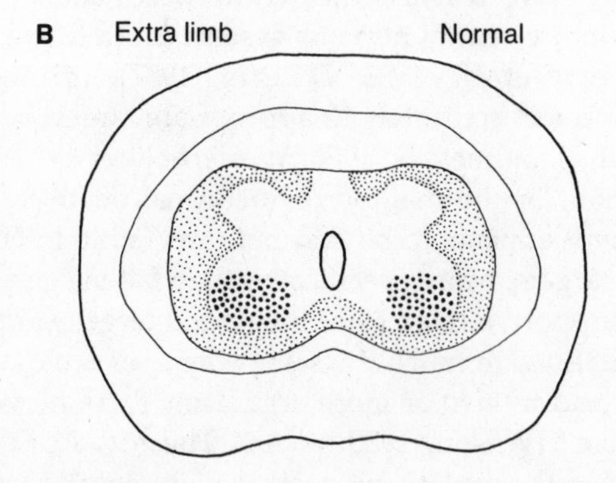

Figure 3.5. Effect of augmenting neural targets on the survival of related neurons. (*A*) Transplantation of a supernumerary limb bud prior to the normal period of cell death in a chick embryo. (*B*) Cross-section of the lumbar spinal cord in a late-stage embryo after limb transplantation. This procedure leads to an abnormally large number of limb motor neurons (dots) on the side related to the extra limb. (After Hamburger, 1977; Hollyday and Hamburger, 1976)

limb in a chick embryo before the period of motor neuron death (Hollyday and Hamburger, 1976; Hollyday et al., 1977). As a result, the number of neurons that survive to maturity is substantially increased. Another way to demonstrate this point is to reduce the size of the original population of neurons so that relatively more target is again made available to a given number of nerve cells (Pilar et al., 1980). In this case as well, the percentage of neurons that survives to maturity is increased. A final experimental example is the increased survival of retinal ganglion cells in mammals after removal of one eye (Chalupa et al., 1984).

A naturally occurring cell salvation of this sort has been described in rodents, in which some muscles are differentially affected by sex hormones (Breedlove, 1986; Fishman and Breedlove, 1987). In adult rats, the bulbocavernosus muscle, which is concerned with the function of the penis, is present in the male but not the female. The muscle is innervated by a small nucleus of motor neurons in the lumbar spinal cord of males, which is absent or rudimentary in adult females (Figure 3.6). In newborn female rats, however, the bulbocavernosus muscle is present and is innervated by a spinal nucleus indistinguishable from that of male rat pups (Sengelaub and Arnold, 1986). The implication of these observations is that the production of androgens in males, but not females, sustains the musculature of the penis, which in turn salvages the related spinal motor neurons from cell death. Support for this interpretation comes from the observation that androgen treatment of newborn females causes them to retain both the male musculature and its innervation (Nordeen et al., 1985). Since a similar sexual dimorphism of the perineal musculature and its innervation is found in most other mammals, including man (Forger and Breedlove, 1986), a shaping of some parts of the nervous system by the action of sex hormones on somatic targets is probably a widespread phenomenon.

Based on evidence of this kind, neurobiologists now accept the idea that the adult number of vertebrate neurons is in many instances the result of an initial excess that is reduced in early embryonic life by trophic interactions at the level of the innervated targets (both somatic targets and targets within the nervous system itself). Because neuronal survival is roughly proportional to the amount of target available, neurons are inferred to compete with one another for a target-derived resource that is available in limited supply. This limitation, and the ensuing competition, are crucial, for if the resource sought by neurons were unlimited, then targets could not influence neuronal numbers as they do.

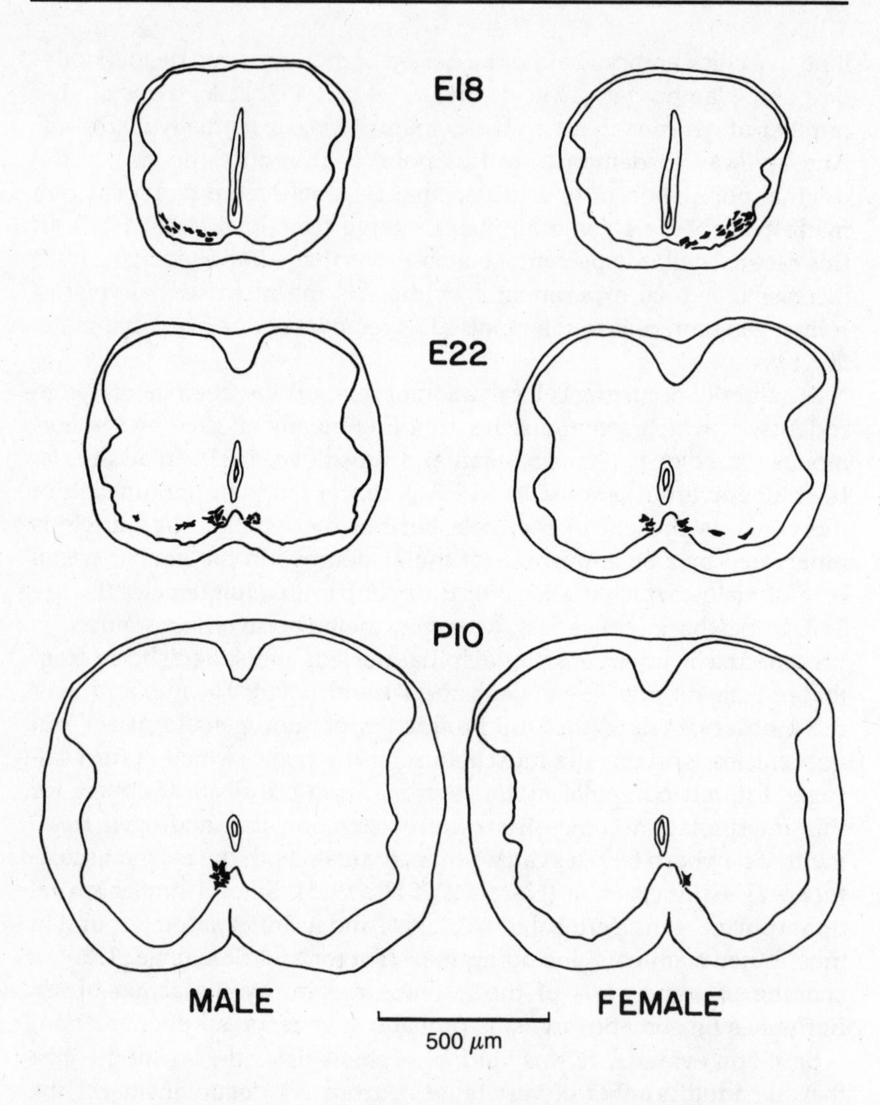

Figure 3.6. Sex-related differences in the number and arrangement of motor neurons in the rat spinal cord. In late embryonic life (embryonic days 18–22), both male and female rats possess a set of neurons that innervates the bulbocavernosus muscle. These nerve cells (shown on opposite sides of the lumbar spinal cord in the two sexes) can be labeled with horseradish peroxidase by retrograde transport after injection into the perineal muscles. In postnatal life, the bulbocavernosus muscle atrophies in the female (because of insufficient levels of male sex hormones); concomitantly, the related motor neurons degenerate. Thus, by the tenth postnatal day (P10), the spinal nucleus of the bulbocavernosus muscle has largely disappeared in the female. The more lateral cells at early stages are thought to be migrating toward the definitive nucleus. (After Sengelaub and Arnold, 1986)

Some Provisos

This account of neuronal death and its basis during the development of vertebrates ignores several complications and uncertainties that should be mentioned. First, neuronal death is only one aspect of the developmental processes that influence the ultimate size of neuronal populations. The factors that determine the size of the *initial* population of nerve cells are not understood at present. Nor is it evident why vertebrate embryos generate a 2- or 3-fold excess of neurons; given the observed somatic variation among individuals of most vertebrate species, a far smaller initial surplus would seem sufficient to allow the necessary adjustments to occur.

Second, the number of surviving neurons is probably not a simple function of the amount of target available (Lamb, 1980; Tanaka and Landmesser, 1986; but see Herrup and Sunter, 1987). For example, if twice the normal number of limb motor neurons is forced to innervate a single limb by early embryonic surgery, the degree of motor neuron death is not simply doubled: considerably more neurons survive than expected on the basis of a simple proportionality between target mass and neuron number (Lamb, 1980, 1981). Nor is there a simple proportionality between the degree of normal neuronal death and target size among related animals of different size. Thus the fraction of the initial population of motor neurons that dies is about 50 percent in various animals that have been examined, even though the ratio of neurons to target cells varies in a disproportionate manner (relatively fewer neurons are available to innervate homologous targets in larger animals—see Chapter 2). Accordingly, the interactions between neurons and targets are likely to be complex, involving, for example, modulation of the trophic properties of target cells by innervation itself (Purves, 1980).

Third, in addition to retrograde influences arising from competition at the level of targets, in at least some circumstances the extent of neuronal death among populations is influenced by the innervation that nerve cells *receive*. Thus, the number of surviving limb motor neurons is also depleted to some degree if the dorsal roots, which carry sensory information to the spinal cord from the limb, are cut prior to the period of normal motor neuron death (Okado and Oppenheim, 1984). A similar increase in cell death is observed in at least some brainstem nuclei (Peusner and Morest, 1977; Parks, 1979) and autonomic ganglia (Furber et al., 1987) if afferent innervation is removed in early embryonic life. These observations suggest that populations are determined by both target-derived and innervation-dependent influences (Chapter 8).

Fourth, there is considerable evidence that the dependence of neurons on their targets varies as a function of age, being profound in early embryonic life and less thereafter (Chapter 7). Moreover, neuronal death and birth persist in maturity in some parts of the vertebrate nervous system, such as the olfactory epithelium (Graziadei and Monti Graziadei, 1978). Finally, the general argument about a trophic basis for neuronal death and survival is based entirely upon observations in vertebrates; whether neurons in various invertebrates obey the same sort of rules is not known. Several studies of this issue suggest that neurons in some invertebrates may not depend on their targets to the same extent as do vertebrate neurons (Sanes et al., 1976; Whitington et al., 1982; Weeks and Truman, 1984, 1985; Arbas and Tolbert, 1986).

Regulation of Neuronal Populations by Their Targets in Ontogeny and Phylogeny

The interactive regulation of neuronal populations in vertebrates according to the size and form of somatic targets (or targets within the nervous system) makes considerable sense from both a developmental and an evolutionary vantage. During development, this refinement allows large and, to some degree, indeterminate numbers of neurons and target cells in large and complex animals to coordinate their associations with a minimum of specific instruction. Whereas a detailed *a priori* program for the fate of every cell is plausible for some very small animals, feedback and consequent adjustment are a necessity when billions or trillions of cells are involved in the gradual construction of an adult animal. A strategy of feedback and adjustment may be equally necessary in evolution to achieve the coordinated changes of body and nervous system that occur from generation to generation. For instance, in the case of land mammals that became aquatic, the numbers of neurons needed to innervate the sort of limbs necessary for terrestrial locomotion had to adjust to the requirements of the different limbs needed for swimming (Figure 3.7). Evidently this sort of adjustment is obvious in the nervous systems of whales: "One of the chief features of the cetacean spinal cord is a reduction of the motor limb nuclei and the almost rudimentary state of the sensory components. Associated with the development of the flippers, there is a prominent cervical swelling of the spinal cord, while a lumbar enlargement is much less prominent due, no doubt, to the vestigial character of the posterior appendages"

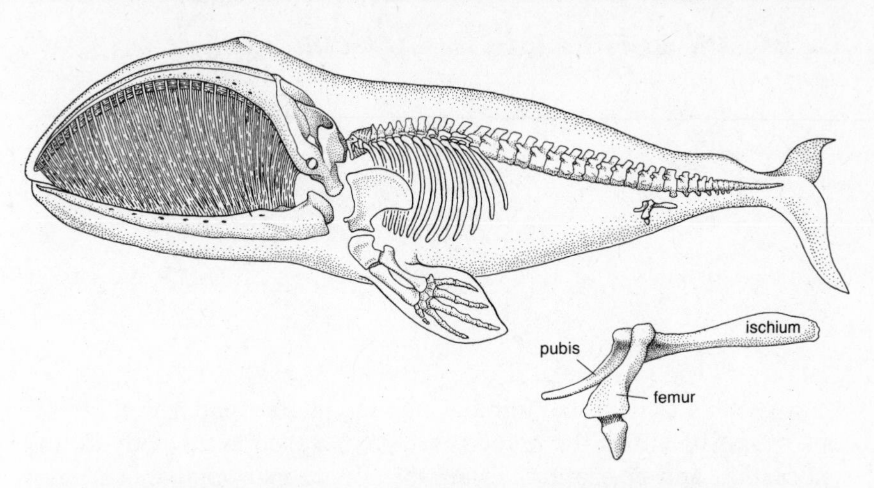

Figure 3.7. Rudimentary hindlimbs of a Greenland right whale (*Balaenoptera mysticetus*). Enlargement of the major bones is shown below. Somatic adaptations of this sort during the course of evolution require commensurate adjustments of neuronal numbers in the corresponding regions of the nervous system, in this case the lumbar spinal cord. Trophic interactions between targets and the neurons that serve them presumably facilitate these changes in the process of speciation, just as they promote neural adjustments to changing size and form during development. (After Romanes, 1895)

(Morgane and Jacobs, 1972, p. 235). The same sort of argument applies to the rudimentary limbs of other vertebrates, such as some snakes and flightless birds. Indeed, this reasoning applies to any adaptation that involves altered neuronal targets.

In summary, trophic interactions between nerve cells and their targets allow the size of neuronal populations in vertebrates to be determined in the course of developmental (or evolutionary) events, as neurons compete with peers striving for success in innervating the same targets. The importance of such interactions for the present argument is that they reveal the nature of the physiological link that exists, in both ontogeny and phylogeny, between an animal's body and the organization of its nervous system.

FOUR

Neuronal Form and Its Consequences

REGULATION of neuronal number through interactions with targets is one means by which the nervous system is yoked to the body during maturation and speciation. Establishing neuronal populations of an appropriate size early in development is, however, an incomplete solution to the problem of neural adjustment to a changing body. First, neuronal populations in mammals and other vertebrates are generally established early in development, when a major part of somatic growth still lies ahead. Second, the size of neuronal populations does not change in proportion to somatic size among related species. These observations imply that the nervous system has at its disposal additional mechanisms of compensation. One such means is modification of neuronal form. By modulating neuronal branching and connectivity, each neuron may innervate few or many target cells (a variable referred to by the term *divergence*); conversely, each target cell may be innervated by few or many axons (a variable referred to by the term *convergence*). Variations in neuronal form, which entail variations in synaptic connectivity, have important functional consequences in both ontogeny and phylogeny.

Diversity of Neuronal Form

The diversity of morphology observed among the neurons of mammals and other vertebrates (both within and across species) is quite different from that observed among cells in other tissues and organs (Figure 4.1). Whereas the majority of somatic cells are roughly similar in size and overall configuration (a nucleus within a cell body that measures a few tens of microns in diameter), nerve cells bear axonal and dendritic arbors that frequently have an aggregate length of millimeters or even meters in larger animals. Thus, by both linear dimensions and volume, neurons may be tens of thousands of times larger than non-neural cells. Moreover, whereas some neurons have extensive axonal and dendritic arbors, others are relatively simple, even when cells from the same part of the nervous system are compared.

Since the introduction of silver staining more than a century ago, neurobiologists have used the morphological diversity of neurons as a means to categorize them. The ability to classify neurons according to shape was greatly enhanced about 20 years ago by the introduction of intracellular marking techniques (Stretton and Kravitz, 1968; Kater and Nicholson, 1973). Because electrophysiological measurements can be made with the same electrode used to inject a marker substance into a particular neuron, this methodology allows neurobiologists to correlate the form of a nerve cell with its function. In spite of these technical advances and their wide application, the significance of neuronal form has remained largely obscure. Although there is no single answer to the puzzle presented by the variability of neuronal form, one important reason for diverse neuronal geometries is apparently to allow neurons to innervate, and be innervated by, more or fewer cells. As in the case of neuronal number, it is useful to explore this idea in relation to animals of different size and complexity in the context of both ontogeny and phylogeny.

Neuronal Form in Small, Simple Animals

Although there are a few instances of functionally unique cells in vertebrates—the Mauthner cell in the brainstem of teleosts, cyclostomes, and some amphibians is the preeminent example—neurons of this sort are exceedingly rare among relatively large, complex animals. In some small, simple invertebrates, however, particular functions are often served by single neurons which are physiologically unique. Neurons that serve the same function in a series of individuals from the same invertebrate species tend to have uniform branching patterns, sometimes allowing particular neuronal branches to be reliably named (Figure 4.2A; Stretton and Kravitz, 1973; Levinthal et al., 1976; White et al., 1976; Truman and Reiss, 1976). Moreover, in some instances stereotyped geometries are maintained in the face of perturbations such as target ablation that have a marked effect on neurons in larger, more complex animals (Figure 4.2B; Sanes et al., 1976; Whitington et al., 1982; Whitington and Seifert, 1982; see also Weeks and Truman, 1984, 1985; Arbas and Tolbert, 1986).

These observations notwithstanding, the geometry of homologous neurons in relatively small, simple invertebrates is neither entirely in-

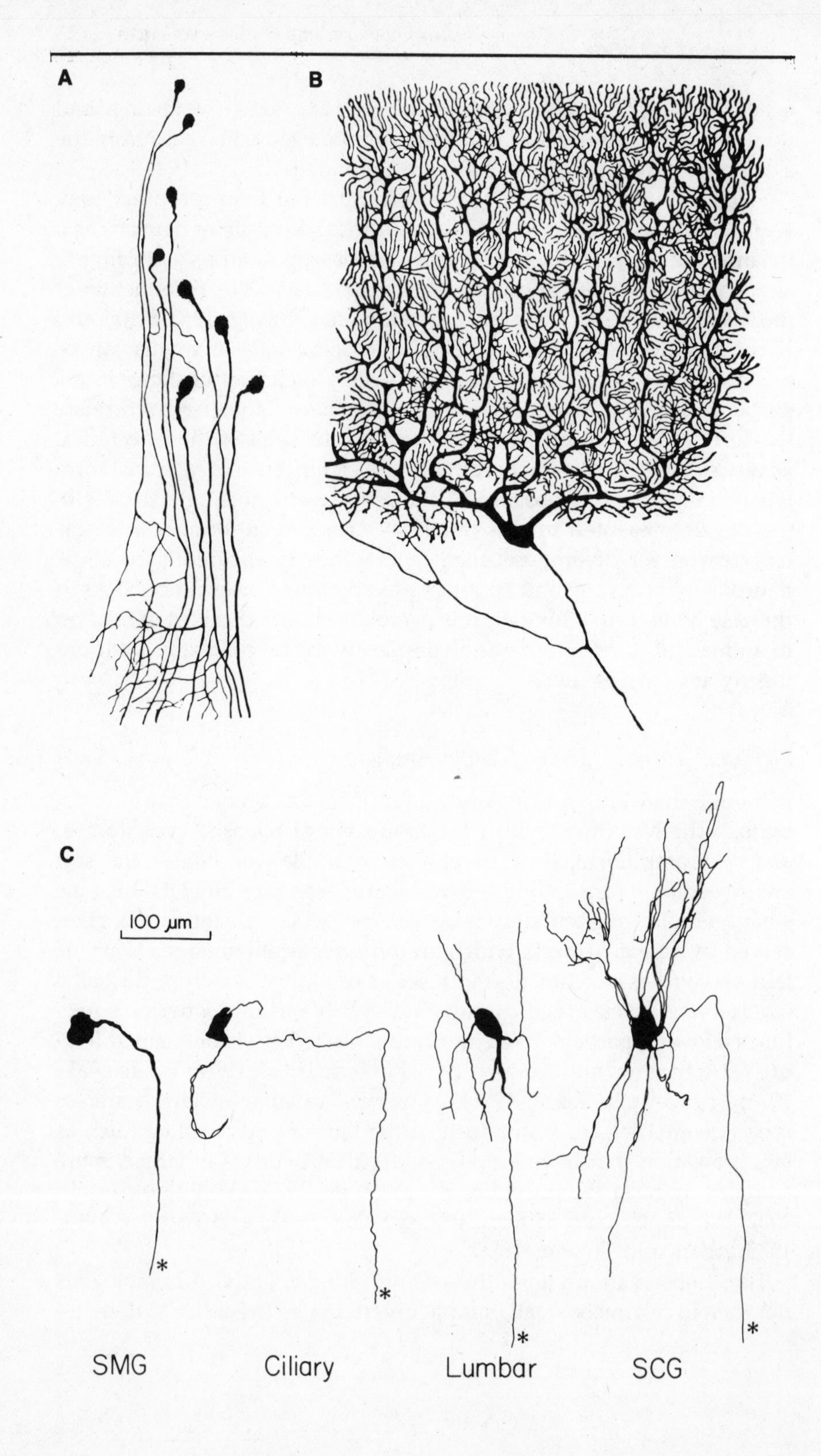
A
B
C
100 µm
*
SMG
Ciliary
Lumbar
SCG

variant nor insensitive to circumstance. First, substantial variability of form among homologous neurons in individuals of the same invertebrate species is apparent at the level of higher order branches. And careful observation of a large number of individuals often reveals variations in the patterns of major branches as well as higher-order neurites (Pearson and Goodman, 1979). Second, just as ablating a particular cell in simple invertebrates can change the fate of neighboring cells, so altering conditions of development can affect neuronal form. In the developing cricket, for example, the axonal arborization of sensory neurons is changed by the presence or absence of neighboring sensory cells (Murphey, 1986a; Shepherd and Murphey, 1986; see also Blackshaw et al., 1982). In the leech, some degree of interaction between neurons and targets is also apparent during development. Thus, a conspicuous pair of neurons found in each segmental ganglion (the Retzius cells) initially develops the same overall form throughout the nervous system. However, Retzius neurons in ganglia that reside in the genital segments begin to acquire a different morphology at about the time their peripheral processes come into contact with the genital primordia, suggesting an influence from these targets (Glover and Mason, 1986; Macagno et al., 1986; Jellies et al., 1987; Loer et al., 1987). Finally, the form of many invertebrate neurons is also influenced by hormonal action (Levine and Truman, 1985; Levine et al., 1986; Levine, 1987). These findings make plain that the form of invertebrate neurons (just like the size of neuronal populations at this taxonomic level) is not generated in a fully determinate manner. Nevertheless, it is fair to say that the detailed form of homologous nerve cells in relatively small, simple animals of the same species tends to be quite similar.

Figure 4.1. Diversity of neuronal form in mammals. (*A*) Some neurons in mammals and other vertebrates are characteristically simple in shape, consisting of only a cell body and an axon. These nerve cells are from the trigeminal nucleus in a mouse. (*B*) Other neurons are strikingly baroque, having hundreds or even thousands of dendritic branches in addition to the axon (which runs off to the bottom right). This neuron is a human cerebellar Purkinje cell. (C) In even the simplest part of the mammalian nervous system, the autonomic ganglia, there is a wide range of neuronal geometries in any given species. These representative cells are from various ganglia in the rat (SMG = submandibular ganglion; SCG = superior cervical ganglion). Asterisks, here and in subsequent figures, indicate the postganglionic axon of each nerve cell; only the portion of the axon near the cell body is shown. The neurons in *A* and *B* were visualized by silver staining; those in C, by intracellular injection of the marker enzyme horseradish peroxidase. (*A* and *B* from Ramón y Cajal, 1911; *C* courtesy of W. D. Snider)

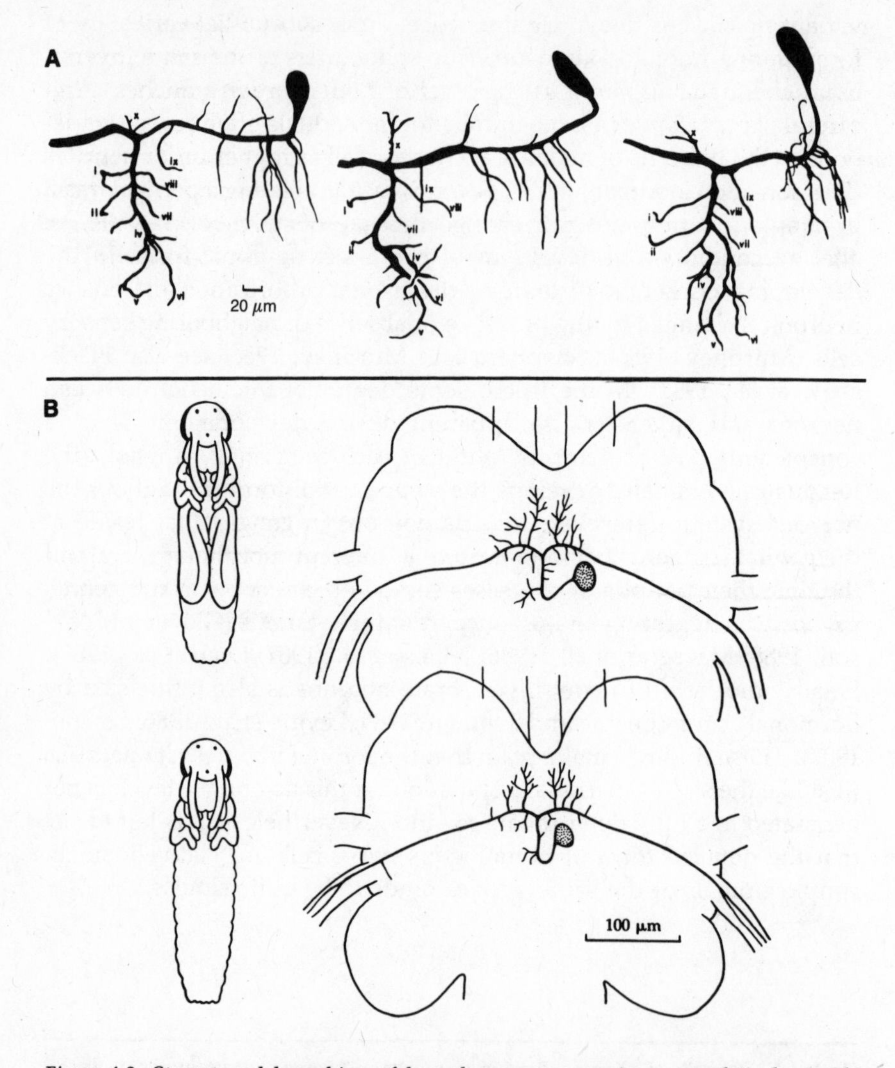

Figure 4.2. Stereotyped branching of homologous neurons in some relatively simple invertebrates. (*A*) An identified motor neuron in three different hawk moths (*Manduca sexta*); the neuron is located by virtue of its characteristic placement and electrophysiological properties. When filled with an intracellular marker (cobalt ions, in this case), the several neurons reveal a remarkably consistent pattern of branching, so that first-, second-, and even some third-order branches can be reliably named (Roman numerals). (*B*) An identified motor neuron in the metathoracic ganglion of (top) a normal grasshopper (*Schistocerca nitens*) and (bottom) a grasshopper after removal of the metathoracic legs at an early embryonic stage. The experimental procedure is shown on left. The motor neurons are subsequently visualized by intracellular injection of a fluorescent dye. Not only do these cells survive in the absence of their normal target, but this circumstance has little or no effect on their geometry. (*A* from Truman and Reiss, 1976; *B* after Whitington et al., 1982)

Neuronal Form in Large, Complex Animals

In contrast to some smaller invertebrates, the morphologies of neurons within the ensembles of cells devoted to a particular function in vertebrates do not show detailed similarities (Figure 4.3). To be sure, the cells of one neuronal type are quite different, on average, from those of another type (compare, for example, the geometries of the neuronal classes in Figure 4.1). But the branches of functionally equivalent cells in vertebrates cannot, as a rule, be individually named: their patterns are far too variable. This variability of neuronal form among neurons of the same class in vertebrate species reflects, in part, axonal and dendritic adjustment to targets which are variable in size and form and which continue to change as such animals mature.

That axonal arbors are regulated by target interactions in vertebrates is apparent from the innervation of skeletal or smooth muscles, autonomic ganglia, or various receptor surfaces, such as the skin. Consider, for example, the innervation of vertebrate skeletal muscle. As an animal matures, muscle mass increases so as to motivate larger skeletal elements. The primary means of muscular growth in vertebrates is by an increase in the length and cross-sectional area of the existing muscle fibers (Ontell and Dunn, 1978; Sperry, 1981; see also Yamauchi and Burnstock, 1969). In some instances, however, substantial numbers of fibers are also added after embryonic development is over (Betz et al.,

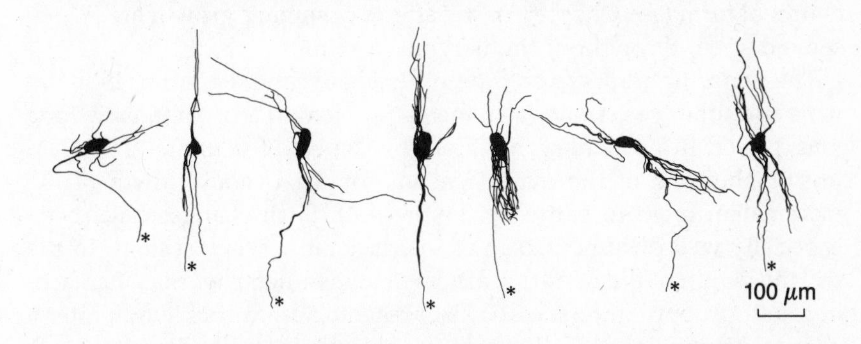

Figure 4.3. Variable morphology of homologous neurons in mammals and other vertebrates. Each of these neurons is from a parasympathetic ganglion that innervates a single target, the salivary gland, in a different rabbit. In contrast to the relatively stereotyped branching of functionally similar neurons in small and simple invertebrates, homologous neurons in mammals and other vertebrates tend to be diverse in shape. The same diversity would be apparent if the neurons had been selected from the salivary ganglion of a single rabbit. (After Snider, 1987)

1979; Harris, 1981; Sperry, 1981; Sassoon and Kelley, 1986; Jones et al., 1987). Whether muscular compensation for growth occurs by fiber growth or fiber addition, a commensurate response of motor neurons must occur. In instances in which more fibers are added, additional axon branches and terminals must be provided to innervate them. In the more general case, in which existing muscle fibers increase in length and girth, motor axons also need to change, because larger muscle fibers require larger synaptic terminals to release enough neurotransmitter to depolarize them to threshold (Kuno et al., 1971). In fact, a progressive increase in the size of motor endplates is observed in vertebrates during the period when the body is growing (Nystrom, 1968a,b; Steinbach, 1981; Hopkins et al., 1985; Steinbach and Bloch, 1986; Lichtman et al., 1987). In some vertebrate muscles, additional endplates, as well as larger terminal arbors at each endplate, are established as muscle fibers exceed a certain length (Bennett and Pettigrew, 1974, 1975; Nudell and Grinnell, 1983). This ongoing addition is also necessary for the innervation of tonic muscle fibers, which normally receive multiple synaptic contacts along their length (Morgan and Proske, 1984). Because vertebrate neurogenesis is, for the most part, complete in embryonic life, the requisite adjustments must be made by ongoing axonal growth of the existing population of motor neurons. The same general arguments can be made for the innervation of viscera by autonomic neurons and the innervation of the sensory surfaces of the body. Furthermore, because increased axonal growth and branching in the periphery are reflected centrally in commensurate changes of neuronal form (see Chapter 6), the effects of somatic growth are felt, in some degree, throughout the nervous system.

The dendritic arbors of both peripheral and central neurons in mammals and other vertebrates also increase in length and complexity for a long period in postembryonic life. This aspect of neuronal growth is presumably one of the major reasons for the overall growth of the mammalian brain in early life (Figure 4.4). Such changes have been especially well documented in the peripheral nervous system. In the rat, the length of the dendritic arbors of sympathetic neurons increases in rough proportion to somatic size, reaching an overall length after a year or more which is 8 or 9 times that observed at birth (Figure 4.5A-B; Voyvodic, 1987b). In man, ganglion cell dendrites apparently increase in complexity over many years (Levi, 1930; Terni, 1930). Neurons in the mammalian brain also show a prolonged postnatal increase in dendritic length, although the extent and duration of this change depends on the species and the region of the brain examined. In the rat, the dendrites of cerebellar Purkinje cells continue to grow for at least a

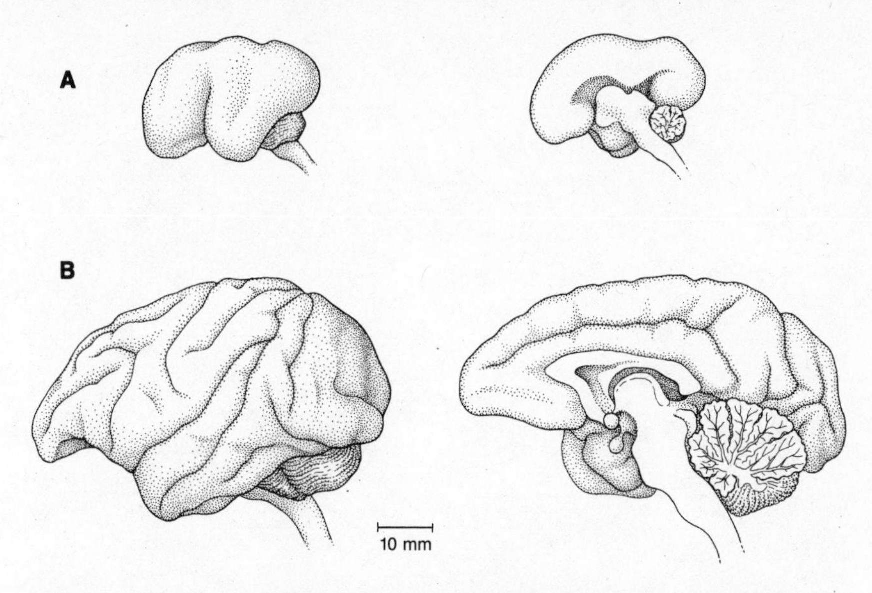

Figure 4.4. Growth of the primate brain in early life. (*A*) Lateral (left) and medial (right) views of a rhesus monkey brain late in gestation (embryonic day 97). (*B*) Similar views of a monkey brain 5 months after birth. Although neurogenesis is virtually complete by embryonic day 100 (see Figure 2.3), the brain enlarges markedly in early life. This observation implies ongoing growth of nerve cells and their processes. (After Goldman-Rakic and Rakic, 1984)

month or two in postnatal life (Figure 4.5C; Addison, 1911; Berry et al., 1980), and the dendrites of mitral cells in the olfactory bulb grow for several months (Hinds and McNelly, 1977). In the rhesus monkey, the dendrites of granule cells in the hippocampus also extend and become more complex for at least some months after birth (Duffy and Rakic, 1983), and in man dendritic growth of cortical neurons may continue for years (Buell and Coleman, 1980). Given the substantial increase in the overall size of the developing mammalian brain without the addition of further nerve cells, increased neuronal branching is hardly surprising.

Because dendrites are a major target for other nerve cells, dendritic growth implies commensurate axonal growth; just as the growth of peripheral targets requires ongoing adjustment of the terminals of primary motor and sensory axons, so dendritic growth requires adjustment of the innervation made by neurons entirely within the nervous system. The fact of axonal and dendritic growth during a prolonged period of maturation must mean the persistence, during this time, of trophic mechanisms that control the orderly growth of neuronal pro-

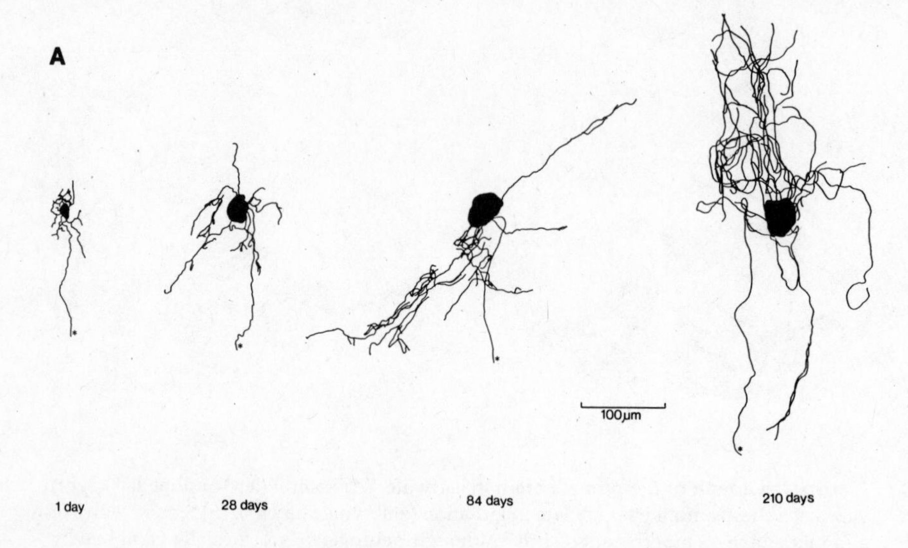
A
1 day
28 days
84 days
210 days
100µm

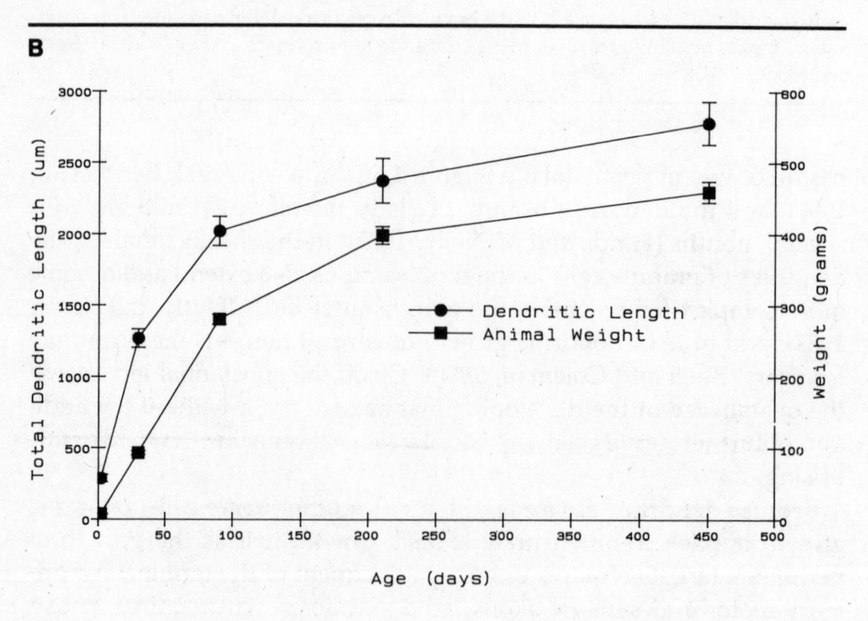
B
Total Dendritic Length (um)
Weight (grams)
Age (days)
Dendritic Length
Animal Weight
0
500
1000
1500
2000
2500
3000
100
200
300
400
500
600
50
150
250
350
450

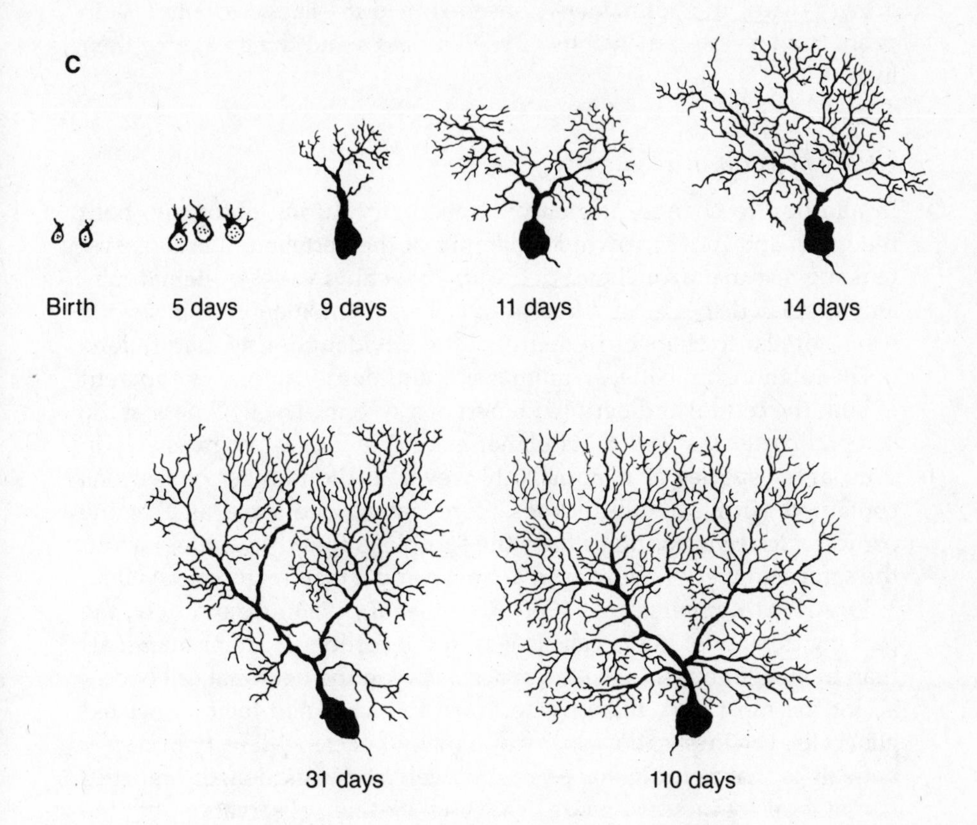

Figure 4.5. Ongoing changes in the length and complexity of neuronal dendrites in the postnatal rat. (*A*) Growth of the dendritic arbors of sympathetic ganglion cells during the first 7 months of life. The neurons were visualized by intracellular injection of horseradish peroxidase. A typical neuron is shown for each age. (*B*) Increase in body weight of the rat as a function of age, together with parallel changes in the overall length of sympathetic ganglion cell dendrites. Each point on the upper curve represents the mean of about 50 neurons from 5–10 animals examined at that age; bars indicate the standard errors. (C) Growth of the dendritic arbors of rat cerebellar Purkinje cells during the first several months of life. The neurons were visualized by silver staining. Again, a typical neuron is shown for each age (in postnatal days). (*A* and *B* after Voyvodic, 1987b; C after Addison, 1911)

cesses and the number and pattern of connections they make. Neurons and their processes also grow in small, simple animals, but this growth—and the adjustment consequent to it—is necessarily less in proportion to the dimensions of such animals and the brevity of their lives.

Neuronal Form in Phylogeny

In addition to changes in neuronal geometry during ontogeny, both the gross and the microscopic anatomy of the vertebrate nervous system suggest that axonal and dendritic branching varies systematically among related species of different size. This phenomenon is, in several ways, similar to changes in neuronal form evident during maturation.

The relationship between animal size and neuronal form is apparent in both the central and peripheral nervous system. The thickness of the cerebral cortex (as well as its planar extent) increases progressively in ever larger species of mammals; however, if the number of neurons contained within a full-thickness core having the same area at the cortical surface is counted, then the sample is found to contain about the same number of neurons in a mouse and a man (Figure 4.6A; Bok, 1959; Ariëns-Kappers et al., 1960; Rockel et al., 1980). Accordingly, the packing density of nerve cells is less in the cortices of larger mammals than in smaller ones. Since the space between the neuronal cell bodies is, for the most part, filled by neuronal branches and their associated glial cells, the implication of the diminished packing density in bigger animals is that neurons have progressively more and longer branches in mammals of increasing size. The long-standing observation that the size of homologous neuronal cell bodies increases across mammals of different body size (Figure 4.6B) is consistent with this idea: neurons with more extensive branches presumably require more protein-synthetic machinery to elaborate and sustain the larger arborizations (Hardesty, 1902; Lapicque, 1946; Bok, 1959). These inferences about neuronal branching have been verified in the few studies in which the dendritic arbors of nominally homologous neurons were measured in the brains of different animals (Figure 4.6C; Bok, 1959; Barasa, 1960; Peichl et al., 1987).

The relationship between the complexity of neuronal branching and animal size is not limited to cells of the cerebral cortex. Studies of the morphology of sympathetic and parasympathetic neurons in homologous ganglia of different mammals also show an increase in the length and complexity of dendrites as a function of species size (Figure 4.7;

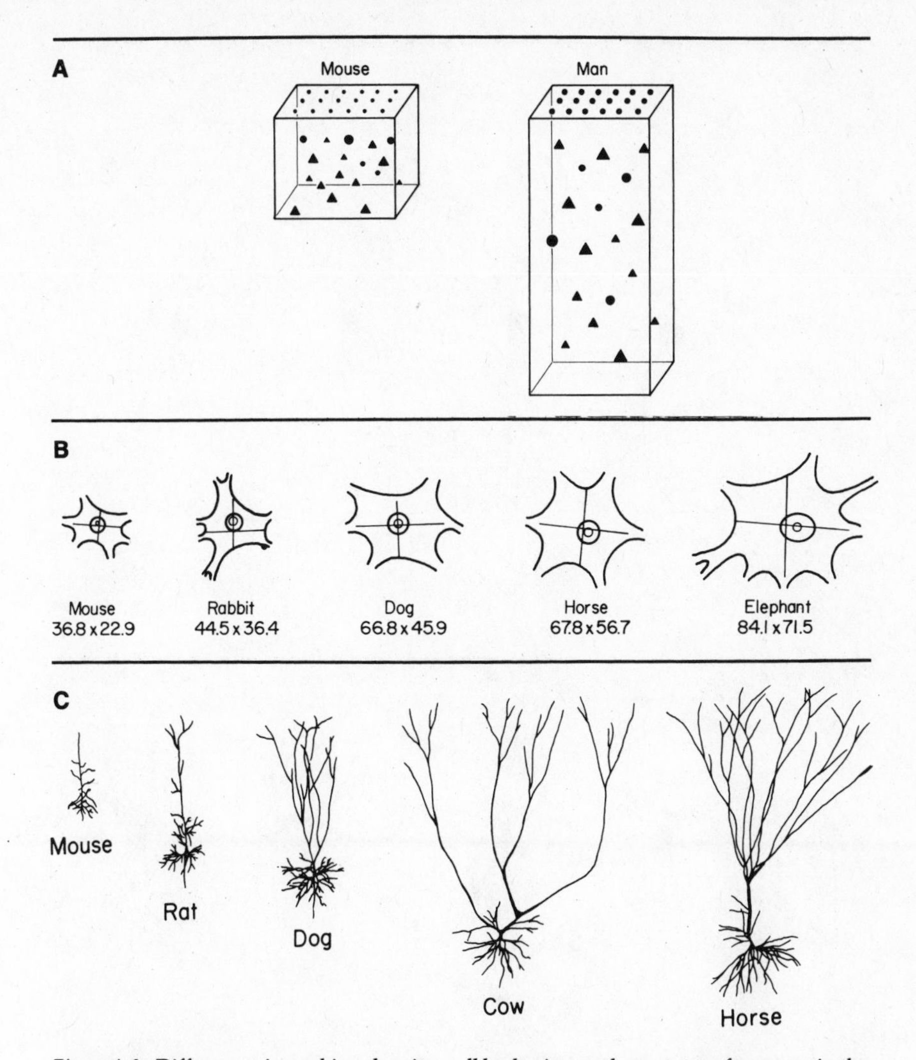

Figure 4.6. Differences in packing density, cell body size, and geometry of neurons in the central nervous system of mammals of different size. (*A*) Density of neurons in the cerebral cortex of a relatively small mammal (mouse) and a relatively large mammal (man). About the same number of nerve cells are counted in a full thickness core of the cortices of the two species. Therefore, the density of neurons in the cortex of man (or other relatively large mammals) is less than the density of neurons in the cortices of relatively smaller mammals. This observation implies that each neuron in a larger animal has more axonal and dendritic branches than the same class of nerve cell in a smaller animal (the different symbols represent homologous neuronal classes). (*B*) Relative sizes of spinal motor neurons measured in histological sections (the average major and minor diameters are shown in microns). The cell bodies of homologous neurons are progressively larger in larger species. This parallel change of neuronal cell body and overall body size presumably reflects the extra synthetic machinery needed to maintain the additional branches elaborated by the neurons in larger species. (C) Typical dendritic arbors of pyramidal neurons from the motor cortices of different mammals. The cells were visualized by silver staining. These observations indicate that the degree of neuronal branching changes systematically in relation to animal size. (*A* after Rockel et al., 1980; *B* after Hardesty, 1902; C after Barasa, 1960)

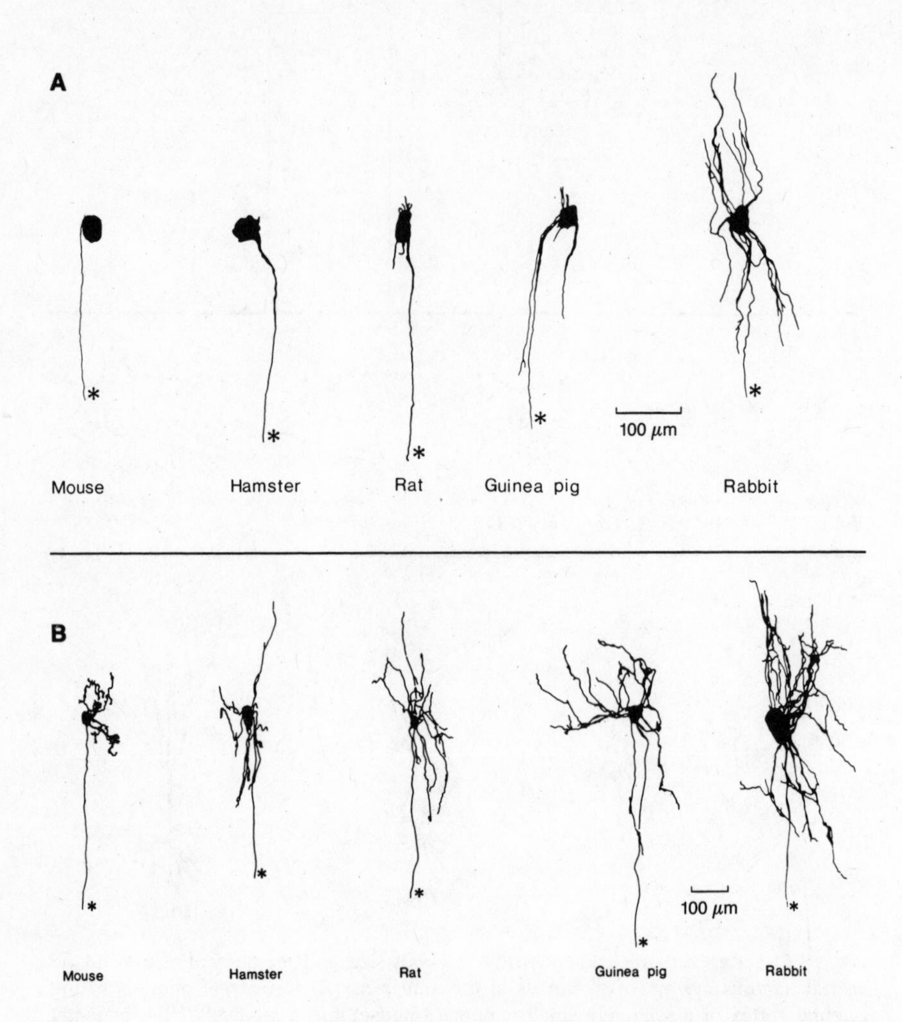

Figure 4.7. Systematic variation of neuronal form among autonomic neurons in mammals of progressively larger size. (*A*) The geometry of representative parasympathetic neurons from the same ganglion (the submandibular; see Figure 4.9A) in several mammals. The cells were visualized by intracellular injection of horseradish peroxidase. (*B*) The geometry of representative sympathetic neurons from the superior cervical ganglion in several mammals. In both parasympathetic and sympathetic divisions of the autonomic nervous system, the overall length and complexity of neuronal branches increase in parallel with the size of the species. (*A* after Snider, 1987; *B* after Purves and Lichtman, 1985a)

Purves and Lichtman, 1985a; Purves, Rubin, et al., 1986; Snider, 1987). These changes in neuronal form in the autonomic system, like those in the central nervous system, are accompanied by changes in synaptic connectivity. Because the density of synapses on the dendrites of autonomic neurons is more or less uniform, the addition of dendritic branches implies the addition of synaptic contacts (Forehand, 1985). Even among autonomic neurons that lack dendrites, the number of synaptic terminals made on the cell body is directly correlated with the overall area of the target cell surface (Sargent, 1983a,b).

The systematic relationship of neuronal morphology and somatic size among related species indicates that regulation of neuronal form is as important in phylogeny as it is in ontogeny. This conclusion has a certain logic: whether the size and form of an animal change during the process of growth or evolution, the nervous system will have to adjust accordingly. Regulation of neuronal form by interaction with the body may be wholly in the service of ontogenetic requirements if the phenotypic variation of the animal has no selective significance; if, however, a particular phenotype happens to be better adapted to the ecological circumstances of that animal, then the same regulatory mechanism can act in the service of speciation.

Functional Consequences of Neuronal Geometry

These morphological differences among neurons in animals of different size and form (whether in ontogeny or in phylogeny) have a variety of functional implications. The consequences of altered axonal geometry are most obviously related to the number of target cells innervated by each neuron (Figure 4.8A). The importance of regulating neuronal divergence can be appreciated by considering the motor requirements of different mammals. Empirically, muscle mass constitutes a constant fraction of the body weight in mammals (about 40–45 percent; Munro, 1969). The homologous muscles in a large mammal (say an elephant) must therefore be larger than those of a smaller mammal (say a mouse). Indeed, this difference should be approximately proportional to the body weights of the two species. However, the numbers of available neurons in both the central and peripheral nervous systems fail to keep abreast of somatic size in larger species (see Chapter 2). Comparative studies have confirmed the disproportionately small number of primary motor neurons in very large mammals (Hardesty, 1902; Tower, 1954; Morgane and Jacobs, 1972). Thus, measurements in the elephant and several more tractable species have shown that the

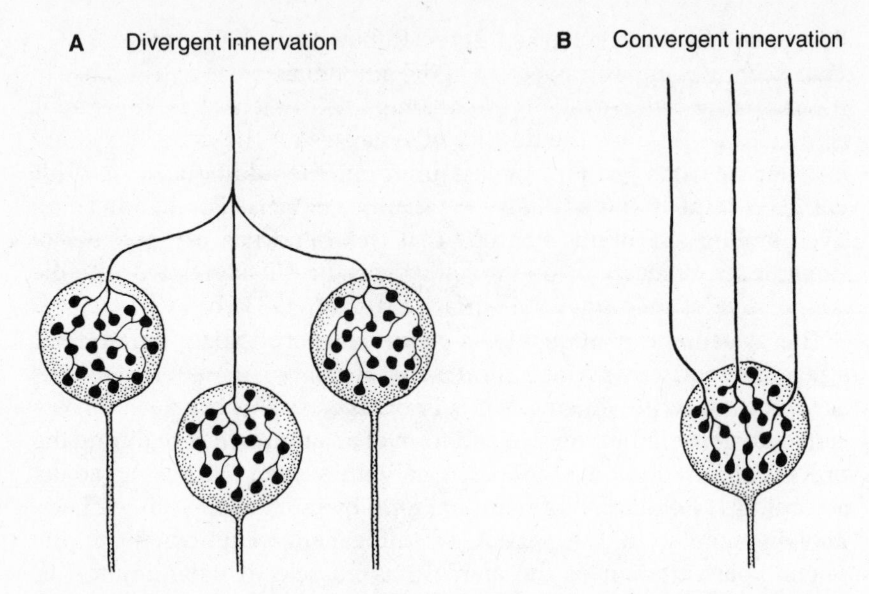

Figure 4.8. Divergent and convergent innervation. (*A*) Each nerve cell axon usually innervates many target cells (which might be effector cells, like muscle fibers, or other nerve cells). This one-to-many relationship is called divergent innervation. (*B*) From the postsynaptic perspective, each target cell may be innervated by many different neurons. This relationship is called convergent innervation. Modulation of divergence and convergence has a major influence on the function of neural pathways.

size of homologous motor centers does not keep pace with body size (for a similar study of whales, see Morgane and Jacobs, 1972). As a result, each motor neuron in an elephant must innervate many more muscle fibers than the homologous motor neurons in a mouse; a corollary is that each motor neuron in a larger animal must support a greater degree of peripheral (axonal) branching. No systematic analysis of the number of muscle fibers innervated by each axon, called the motor unit size, has been carried out in homologous muscles of different mammals. However, studies in which motor unit size has been incidentally determined in the same muscle of different mammals supports this conclusion (for a comparison of motor unit size in the soleus muscle of mouse, rat, and rabbit, see Brown et al., 1976; Bixby and Van Essen, 1979; Fladby, 1987; for values in the medial gastrocnemius muscle of cat, monkey, and man, see Burke, 1981). The same general argument applies to the motor innervation of smooth muscles, for which the relative number of related neurons also falls progressively as animal size increases (see Table 2.2), and to the sensory innervation of muscles and other organs.

One consequence of larger motor units in the homologous muscles of larger animals is that, because of increased divergence, an action potential in a skeletal or visceral motor neuron in a large animal generates more force in the muscle it innervates than an action potential in a small animal. The "power gain" achieved by increased axonal branching at the level of peripheral targets can also be accomplished in the more proximal portions of a pathway; divergence en route to a muscle further increases the power amplification of the pathway. For example, the number of nerve cells innervated by each presynaptic axon in sympathetic ganglia (called the neural unit size, by analogy with motor units) also increases systematically in proportion to animal size (Table 4.1; Purves, Rubin et al., 1986). Whereas each visceral motor neuron in the mouse spinal cord innervates about 60 neurons in the superior cervical ganglion, each homologous preganglionic neuron in the rabbit innervates about 400 ganglion cells (since these determinations require electrophysiological measurements from ganglion cells while stimulating the motor outflow from the spinal cord, such experiments in much larger animals are not practical). As in the skeletal motor system, each neuron in the superior cervical ganglion must diverge to innervate

Table 4.1. The number of superior cervical ganglion cells innervated by each preganglionic neuron (called the neural unit size) in various mammals. The neural unit size is estimated by multiplying the ratio of ganglion cells to preganglionic cells (Table 2.2) by the average number of preganglionic axons innervating each ganglion cell. The number of ganglion cells contacted by each central axon increases in parallel with animal size. The *effective* difference in neural unit size among these animals is less than it appears because not all of the cells in the neural units of the larger animals are brought to threshold by each of the axons that contacts them. (From Purves, Rubin, et al., 1986)

Animal	Approximate weight (g)	Mean no. of preganglionic neurons innervating each ganglion cell	Approximate ratio of ganglion cells to preganglionic neurons	Estimated neural unit size
Mouse	25	4.5	14.2	64
Hamster	100	7.2	25.0	180
Rat	200	8.7	27.2	237
Guinea pig	400	12.3	26.8	330
Rabbit	1,600	15.5	27.2	422

more smooth muscle cells in the head and neck of progressively larger animals. The increased size of neural units in autonomic ganglia further enhances the ultimate influence of each central autonomic neuron.

The functional consequences of varying divergence in these motor pathways go beyond the primary purposes of assuring that each of the muscle fibers (or neurons) in a target receives innervation and that each central neuron generates an appropriate amount of motor power. Again, the innervation of skeletal muscle provides a ready example. The *average* number of fibers innervated by each axon in a given species of mammal varies greatly (Burke, 1981). In general, large muscles comprise motor units that are relatively large, whereas small muscles have smaller motor units. One rationale for this relationship is that the action of a large muscle (for example, a leg muscle like the gastrocnemius) does not require the finely graded response of a small muscle (for example, a lumbrical muscle in the hand). Yet another way that modulation of divergence affects function is by providing a diversity of motor unit sizes within a given muscle, be it the gastrocnemius or a lumbrical. Spinal motor neurons that have relatively small cell bodies innervate a relatively small number of fibers in their target muscle; and these neurons are the first to respond to a sensory stimulus that elicits a reflex motor response. The larger motor neurons, which innervate a proportionally greater number of fibers in the muscle, have higher thresholds. This rule is referred to as the "size principle" (Henneman et al., 1965; Henneman and Mendell, 1981). Because smaller motor units are recruited before larger ones, a given muscle is capable of finely graded movements, as well as coarser ones. These facts give some indication of the intricate problems addressed by the regulation of the number of target cells innervated by each axon, even in a relatively simple system such as the innervation of muscle.

Changing the dendritic geometry of neurons also has a number of functional consequences. An effect that is as important for the organization of neural pathways as divergence is the determination of the number of axons that innervate each nerve cell (that is, convergence) (Figure 4.8B). Evidence that convergence depends on the geometry of the postsynaptic cell has come from studies of the peripheral autonomic system, where it is possible to stimulate the full complement of axons innervating an autonomic ganglion and its constituent neurons; this approach is not usually feasible in the central nervous system of mammals, simply because of the anatomical complexity of most central pathways. Since individual neurons can also be labeled via the intracellular recording electrode, electrophysiological measurements of the number of different axons innervating a neuron can routinely be correlated with cell shape (Figure 4.9). In both the parasympathetic and sympathetic ganglia of various mammals, the degree of preganglionic convergence onto a neuron is proportional to its dendritic complexity (Purves and Hume, 1981; Purves and Lichtman, 1985a; Purves, Rubin, et al., 1986; Snider, 1987). Thus, geometrically simple neurons that lack dendrites altogether (see Figures 4.1C and 4.7A) are generally innervated by a single input. Conversely, neurons with increasingly complex dendritic arbors are innervated by a proportionally greater number of different axons. This correlation of neuronal shape and input number holds within a single ganglion, among different ganglia in a single species, and among homologous ganglia in a range of species.

The covariance of dendritic geometry and convergence leads to a number of effects. Because convergence allows the set of target cells innervated by a particular axon to be innervated by one or more additional axons, a more subtle activation of the target population is possible than when each target cell is activated by a single input. Whereas the activity of a singly innervated target cell can only reflect the electrical activity of a particular presynaptic axon, the activity of a cell innervated by multiple axons will be determined by the entire population of neurons that impinge upon it. As might be expected, the normal pattern of ongoing electrical activity in autonomic neurons is also influenced by the number of inputs they receive, which is modulated in turn by dendritic geometry (Johnson and Purves, 1983). Thus the frequency of electrical activity in sympathetic ganglion cells of a small mammal like the mouse, which have few dendrites (Figure 4.7B), is much less than the ongoing rate of activity in the ganglion cells of a larger animal like the rabbit, which have more complex dendritic arbors (A. Ivanov and D. Purves, unpublished). Multiple innervation of individual target cells also allows inputs to be differentially distributed on the postsynaptic surface. Because of the electrical properties of excitable cells, inputs placed far from the zone where a postsynaptic action potential is initiated are, in general, less effective than inputs placed closer to this zone (Rall, 1959; but see Eccles, 1960; Redman, 1986). Thus, differential location of inputs on dendritic arbors influences the integrating properties of the postsynaptic cell. Finally, the elaboration of progressively more dendrites among the neurons of a target population influences the size of neural units (see Table 4.1 and Chapter 5). Rather than making many endings on a few adendritic cells, each presynaptic axon

A

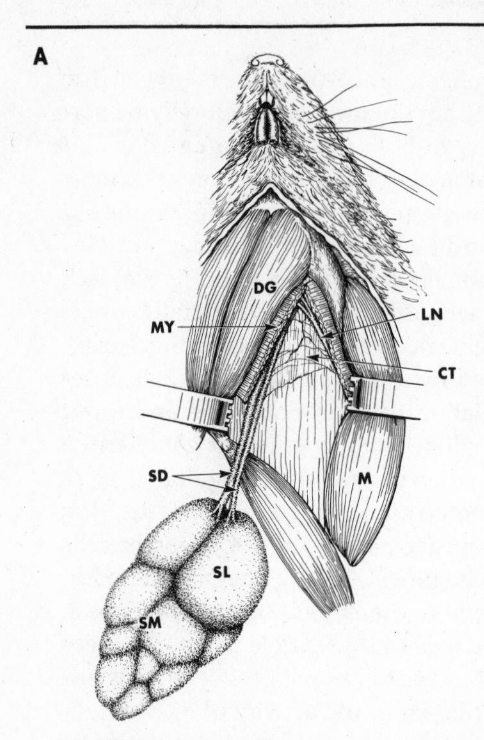
DG
MY
LN
CT
SD
M
SL
SM

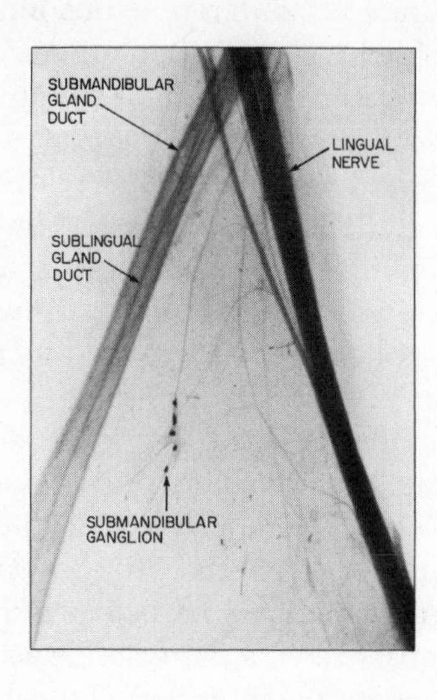
SUBMANDIBULAR GLAND DUCT
LINGUAL NERVE
SUBLINGUAL GLAND DUCT
SUBMANDIBULAR GANGLION

B

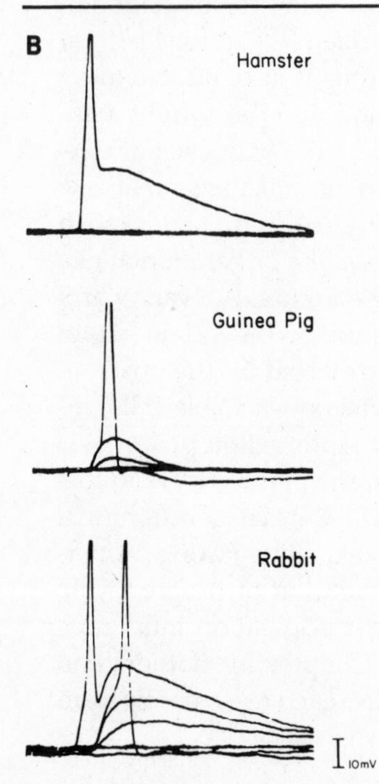
Hamster
Guinea Pig
Rabbit
10mV

C

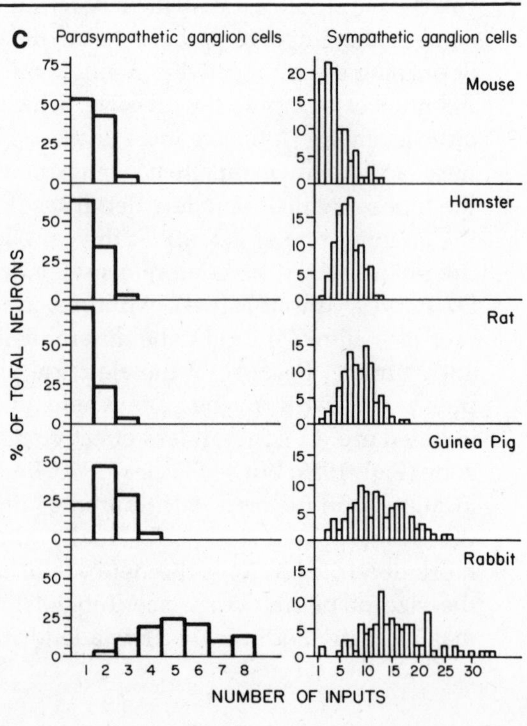
Parasympathetic ganglion cells
Sympathetic ganglion cells
Mouse
Hamster
Rat
Guinea Pig
Rabbit
% OF TOTAL NEURONS
NUMBER OF INPUTS

tends to make a smaller number of endings on a larger number of cells with complex dendritic arbors. In sum, modulation of dendritic form changes the number of inputs that neurons receive, the character of their activity, the manner in which they integrate the synaptic information that impinges on them, and the size of the neural units in which they participate.

In virtually all animals that have nervous systems, most neurons have axonal and dendritic arborizations. There is an important difference, however, in the manner in which the form of neurons is established in various animals. In some smaller invertebrates, the shape of neurons is largely a consequence of determinate developmental programs (bearing in mind that the development of even the simplest nervous system is not fully determinate). As a result, the form of homologous neurons among individuals of the same species tends to be stereotyped. In contrast, homologous neurons in large and complex animals like vertebrates vary greatly in their form, as might be expected from the interactive nature of the development of such animals. Vertebrate neurons do, of course, have a determinate aspect; thus, different classes of nerve cells in a given species have common features that are likely to be genetic in origin. If, for example, vertebrate neurons are removed at an early stage and placed in tissue culture, they often develop some of the features they possess *in situ* (Dichter, 1978; Banker and Cowan, 1979; Wakshull et al., 1979; Kriegstein and

Figure 4.9. Correlation of dendritic geometry and convergent innervation. (*A*) The submandibular ganglion in the rat. This parasympathetic ganglion is located in a connective sheath (CT) that extends between the salivary ducts (SD) and the lingual nerve (LN). MY, DG, and M indicate various muscles; SL and SM indicate the sublingual and submandibular portions of the salivary gland (an actual ganglion appears in more detail in the inset). The number of inputs to a particular cell can be determined by recording from individual neurons with a microelectrode while stimulating the preganglionic (lingual) nerve. (*B*) Representative recordings in 3 of the 5 species studied comparatively. As the strength of preganglionic stimulation increases, increments in the postsynaptic response can be counted (several traces are superimposed in each panel). Each increment corresponds to the recruitment of an additional axon contacting the cell. Neurons in the ganglion of a smaller animal, like the hamster, are usually innervated by a single input, whereas neurons in progressively larger animals are innervated by progressively more axons. (*C*) Histograms of the number of inputs to neurons examined in this way from the submandibular ganglion (left) and the superior cervical ganglion (right) in 5 different mammals. The number of axons innervating homologous neurons increases sytematically as a function of dendritic complexity, which in turn correlates with species size. (*A* from Lichtman, 1977; Snider, 1987; *B* from Snider, 1987; *C* after Purves and Lichtman, 1985a).

Dichter, 1983). These intrinsic contributions to neuronal form notwithstanding, axons and dendrites of vertebrate nerve cells establish many aspects of their shape according to local circumstances. One rationale for this relation is the adaptability that modulation of neuronal form provides to animals in which the numbers of nerve and target cells are not precisely determined, and whose bodies continue to grow long after the sizes of neuronal populations are fixed.

FIVE

Regulation of Developing Neural Connections

INFORMATION gleaned about the establishment of neuronal populations and about the development and significance of neuronal form in vertebrates indicates that the ontogeny of large and complex animals entails a good deal of adjustment. In the case of neuronal populations, the dependence of developing nerve cells on trophic support from their targets has shown that cellular interaction is an important mechanism in the regulation of neuronal number. The factors that underlie the diversity and plasticity of neuronal form have been less certain; only recently has it become clear that the branches of neurons are also subject to trophic regulation through interactions between neurons and the cells they innervate. Evidence about the nature and purpose of the regulation of neuronal branches and their synaptic connections has come, in the main, from two sources: studies of the formation of neural connections during development and studies of the responses of mature patterns of neural connections to experimental perturbation. This chapter focuses on the regulation of neuronal branches and connections during development; the following chapter examines the plasticity of neural connections in maturity. In the final analysis, the same regulatory mechanisms appear to operate, in varying degrees, throughout life.

Formation of Neuronal Connections in Vertebrates

The formation of neural circuits in mammals and other vertebrates extends far beyond embryogenesis. When synapses are counted by conventional methods in tissue sections, the overall number of synaptic terminals in various regions of the nervous system indicates that synaptogenesis continues for weeks, months, or longer (Figure 5.1). Even in the parts of the mammalian nervous system that are simplest in organization (skeletal muscles and autonomic ganglia) the deposi-

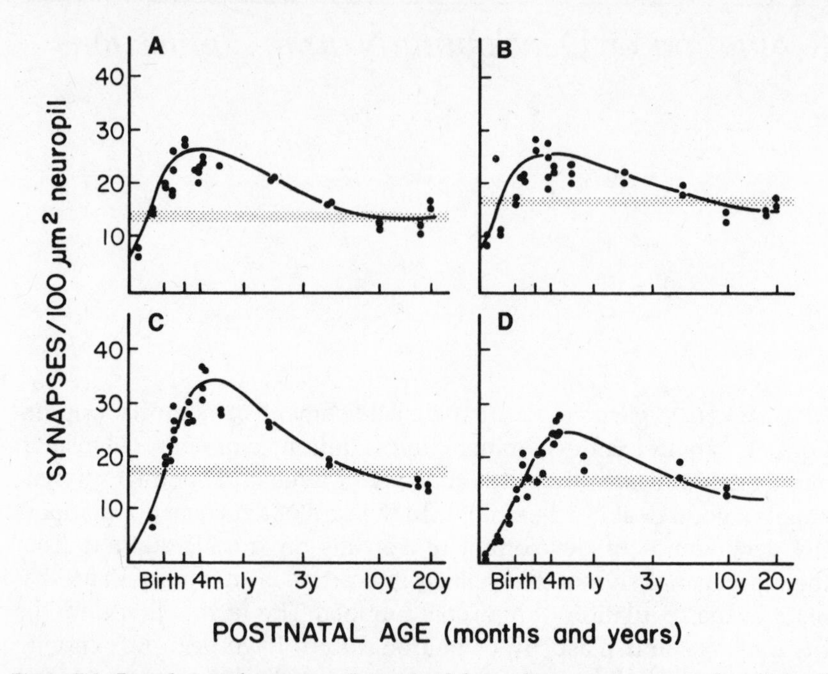

Figure 5.1. Prevalence of synapses determined from electron microscopical sections of the cerebral cortex of the rhesus monkey at different postnatal ages. The regions of cortex examined are primary motor (*A*), primary somatosensory (*B*), prefrontal (*C*), and primary visual (*D*). The stippled lines indicate the average synaptic density in adult animals. Synapses are evidently being formed in large numbers during at least the first several months of a monkey's life. The gradual fall in the number of synapses per unit area of electron microscopical sections after about 6 months of age does not necessarily indicate a cessation of synapse formation. (After Rakic et al., 1986).

tion of synapses is prolonged (Dennis et al., 1981; Smolen, 1981; Rubin, 1985c). Of course, synapse counts do not necessarily indicate when the establishment of additional nerve terminals ceases in maturing animals. Ongoing synaptogenesis could occur without any change in the overall number of contacts counted in histological sections if the rate of synapse formation were counterbalanced by synaptic withdrawal. Therefore, these observations provide a minimum estimate of the duration of synapse formation. In fact, this process probably persists throughout life (see Chapter 6).

The period of net synaptic accretion appears to be longer in larger species. In the rat cerebral cortex, for example, the number of synapses observed per unit area (or volume) of brain increases substantially during the first several weeks of postnatal life (Vrensen et al., 1977;

DeGroot and Vrensen, 1978; Blue and Parnavelas, 1983). In the cortex of the cat, synapse counts increase for about 2 months (Cragg, 1975; Winfield, 1983). In a variety of cortical areas examined in monkeys, the density of synapses increases for several months postnatally (Lund et al., 1977; O'Kusky and Colonnier, 1982; Garey and de Courten, 1983; Rakic et al., 1986). In human visual cortex, synaptic numbers increase for about a year after birth (Huttenlocher, 1979; Huttenlocher et al., 1982; Garey and de Courten, 1983; Huttenlocher and de Courten, 1987).

The gradual elaboration of synapses in mammals and its prolongation in larger species are consistent with, and presumably related to, the gradual growth of axonal and dendritic branches. These anatomical observations also accord with the general idea that patterns of neural connections remain flexible during the period when developing animals undergo obvious changes in size and form.

Synaptic Rearrangement in the Peripheral Nervous System

A deeper understanding of the significance of a prolonged period of synaptogenesis in mammals and other vertebrates has been gained by studies carried out in the peripheral nervous system. The advantage of examining muscle and autonomic ganglia in this regard is, again, the anatomical simplicity of these regions of the nervous system; because it is possible to stimulate the full complement of inputs to such targets, the innervation of individual cells can be assessed quantitatively over time.

Most skeletal muscle fibers in adult mammals are innervated at a single endplate site by one motor axon, although some twitch fibers and all tonic fibers are multiply innervated. In 1970, P. A. Redfern rediscovered by means of electrophysiology a fact known to anatomists earlier in the century (Tello, 1917; Ramón y Cajal, 1925; Boeke, 1932), namely, that neonatal muscle fibers in mammals are usually innervated by terminals arising from several different axons; this relationship is generally referred to as polyneuronal innervation of target cells (Figure 5.2). Studies carried out in the 1970s showed that in mammals like the rat the transition from polyneuronal to the mature one-on-one relationship of motor axons and muscle fibers occurs gradually in postnatal life (Brown et al., 1976; Brown et al., 1981a; Van Essen, 1982; Bennett, 1983). Although mammalian motor neurons may sometimes die during this period (Rootman et al., 1981; Baulac and Meininger, 1983; Bennett et al., 1983), the elimination of a portion of the initial

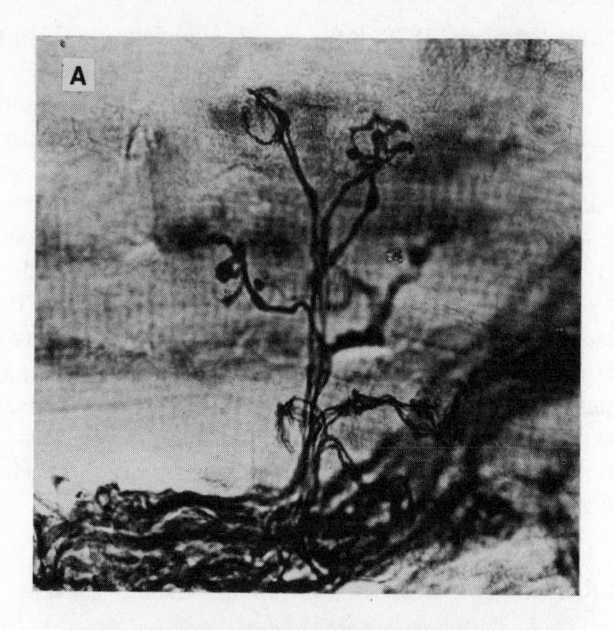

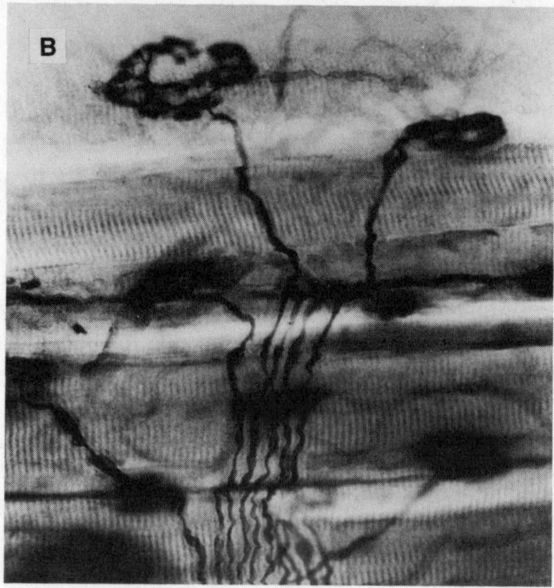

Figure 5.2. Innervation of mammalian skeletal muscle fibers by a greater number of axons at birth than in maturity. (*A*) Silver-impregnated nerve fibers in a leg muscle from a 10-day-old rat. Each of the 3 endplates in the field is innervated by at least 2 different axons. (*B*). In an adult rat (or other mammal), the endplates on skeletal muscle fibers are usually innervated by a single axon, as in this example. The terminal in sharp focus is also more complex than the endings on neonatal fibers. (Courtesy of D. Riley)

innervation of muscle can clearly occur without the loss of any of the spinal motor neurons involved (Brown et al., 1976; Lance-Jones, 1982; Hardman and Brown, 1985; Oppenheim, 1986; Fladby, 1987). Thus the elimination of a portion of the initial innervation to muscles is not a result of ongoing motor neuron death.

A similar phenomenon was discovered among peripheral neurons when neonatal innervation was explored in mammalian autonomic ganglia (Lichtman, 1977; Lichtman and Purves, 1980; Hume and Purves, 1981; Johnson and Purves, 1981). Among ganglion cells that, in maturity, are innervated by a single preganglionic axon, polyneuronal innervation is also apparent at birth. As in mammalian muscle, the number of axons innervating each target neuron declines over the first few weeks of life (Figure 5.3) and this decline is not caused by the death of presynaptic cells (Johnson and Purves, 1981).

One potentially confusing point in a comparison of muscle and ganglia is that the total number of synaptic terminals on ganglion cells actually *increases* during the time that polyneuronal innervation of the target cells decreases (Lichtman, 1977; Johnson and Purves, 1981)—an observation consistent with the gradual increase in the numbers of synapses counted in other regions of the nervous system during early postnatal life. A casual consideration of the development of muscle innervation, however, suggests that a net *loss* of synapses occurs during this period. Appearances to the contrary, there is no real discrepancy between the strategy of innervation among muscle fibers and nerve cells. The manifest reduction in the number of axonal *inputs* to individual muscle cells has sometimes been taken to imply the depletion of an initial surfeit of synaptic terminals (in much the same way that neuronal death in normal development represents the loss of an initial excess). However, the complexity of terminal branching and the number of synaptic sites within each endplate on mammalian muscle fibers increase during the period of input loss in muscle (see Figures 5.2–5.3). In this light, both muscle cells and peripheral neurons show a net *gain* of synapses during early life.

Additional confusion arises because of the obviously different anatomy of muscle cells and neurons. Thus the neuromuscular junction is often referred to as a "synapse" and is therefore equated with a synaptic bouton on the surface of a neuron. However, the nature of synapse formation in muscle and ganglia suggests that the appropriate equivalence is between the endplate site on a skeletal muscle fiber and the entire cell body of a ganglion cell that lacks dendrites (Figure 5.4). Both the endplate and the cell body of an adendritic neuron are preferred sites of innervation of about the same dimensions. And both muscle and ganglion cells reject innervation outside these regions. Thus, neither the majority of the muscle fiber surface nor the axon of the nerve cell are innervated. In spite of apparent differences, the elimination of polyneuronal innervation in muscles and ganglia proba-

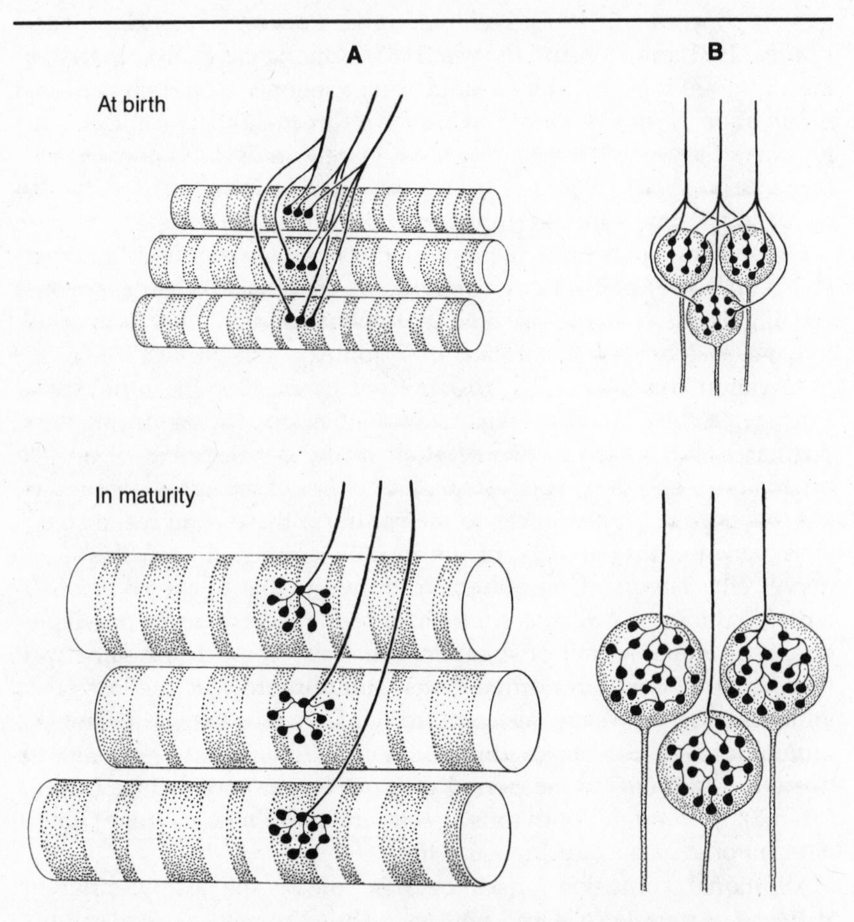

Figure 5.3. Major features of synaptic rearrangement during the first few weeks of post-natal life in the peripheral nervous system of mammals. In muscles (*A*) and ganglia comprising neurons without dendrites (*B*), each axon innervates more target cells at birth than in maturity. In both muscles and ganglia, the size and complexity of the terminal arbor on each target cell increase; thus, during early life, each axon elaborates more and more terminal branches and synaptic endings on the target cells it will innervate in maturity. Accordingly, the common denominator of this process is not a net loss of synapses but the focusing of an increasing amount of synaptic machinery on fewer target cells. (After Purves and Lichtman, 1980)

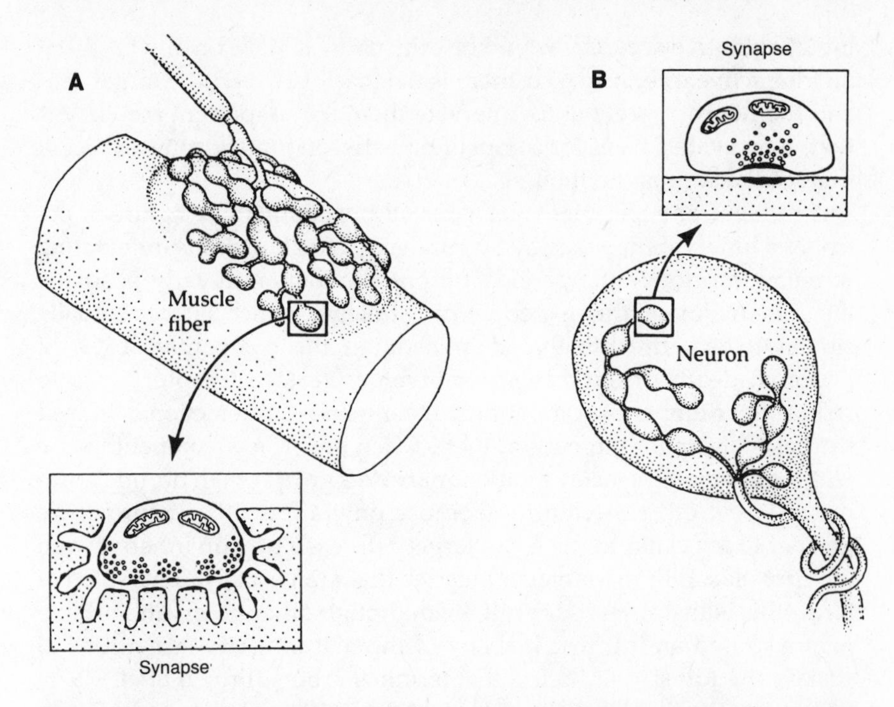

Figure 5.4. Size and arrangement of (*A*) a mature vertebrate neuromuscular junction (from a snake muscle) and (*B*) a mature neuron that lacks dendrites (from a frog ganglion). The blowups of individual boutons indicate that the presynaptic endings are quite similar on the two types of cells. This juxtaposition suggests that the innervation of an endplate is comparable to the innervation of a neuron that lacks dendrites. (After Kuffler and Yoshikami, 1975 and based on drawings by U. J. McMahan).

bly represents a similar process. Failure to recognize this underlying similarity has made comparison of early synaptic rearrangement at different sites more difficult than it needs to be.

Competitive Nature of Synaptic Rearrangement

The mechanism of the transition from multiple to single innervation is generally considered to involve neuronal competition, defined as the dependence of neurons, or in this case their branches, on a limited resource that peers are also seeking to acquire. Several aspects of the development of peripheral innervation support this view. The end result of postnatal input elimination in muscles and ganglia is that every target cell receives an appropriate number of inputs and, *ipso facto*, no cell in the target population is left uninnervated. Such unfailing quantitative accuracy would be difficult to achieve on any basis but an interactive one, in which there is feedback between the target cells and the neurons seeking to innervate them. An important corollary is that, to provide a basis for competition between inputs, some aspect of this feedback must be limiting.

A number of experiments have tested the competitive nature of peripheral innervation explicitly by removing a fraction of the innervating population prior to the period of input elimination in early postnatal life. The point of this experimental strategy, which is conceptually similar to experiments that demonstrated the competitive basis of neuronal death or survival in embryonic life (see Chapter 3), is to reduce the number of competitors. If competitive interaction is indeed the basis of input elimination, then, when only a few competitors are left, each remaining axon should innervate a greater than normal number of target cells in maturity (because only a few other neurons are present to lay claim to the set of target cells each neuron innervated in the first place). In mammalian muscles that are partially denervated at birth, this is the observed result, even though other observations have demonstrated an intrinsic inability of the maturing motor neurons to sustain the full size of their initial terminal arbors (Brown et al., 1976; Betz et al., 1979, 1980; Ribchester and Taxt, 1983; Fladby and Jansen, 1987; see also Heathcote and Sargent, 1985).

Additional evidence of the competitive nature of synaptic rearrangement on developing nerve cells comes from the observation that the adult number of inputs to autonomic ganglion cells is correlated with the geometry of each target neuron (Hume and Purves, 1981; Purves and Hume, 1981). In general, ganglion cells that lack dendrites are innervated in maturity by a single axon, whereas neurons that bear progressively more dendrites are contacted by an increasing number of inputs (Figure 5.5); in fact, up to tens of thousands of inputs may innervate neurons with extensive dendritic arbors. A similar relationship is observed when homologous neurons with different average geometries are compared across species (Figure 4.9). The correlation of neuronal geometry and convergence is probably explained by the effect that dendrites have on the process of input elimination. At birth, all ganglion cells are innervated by a relatively large number of inputs. As the animal matures, there is a reduction of inputs to target neurons in those instances where the nerve cells lack significant numbers of dendritic processes. However, this phenomenon is less and less apparent among neurons that possess increasingly complex dendritic arbors (Figure 5.6). For example, among neurons in the rabbit ciliary ganglion (the only part of the nervous system that has been systematically examined in this regard), the full range of neuronal geometries is present at birth (Hume and Purves, 1981) (although the overall extent of each neonatal cell's dendrites is much smaller than in adult animals, as is expected from the gradual postnatal growth of dendrites; see Chapter 4). Evidently, the vigor of the competitive interactions that cause some initial inputs to be lost from a target cell is reduced by the elaboration of dendrites. Why might this be? Among neurons that lack dendrites, the terminals of the several axons that initially innervate each cell are confined to a restricted arena—the cell body and the proximal part of the postsynaptic axon. In contrast, the axon terminals on cells that possess dendrites are deployed over a surface that is proportional to the geometrical complexity of the postsynaptic cell. A possible explanation of the adult correlation between inputs and geometry, therefore, is that postsynaptic geometry modulates convergence by altering the arena for competitive interaction among axon terminals innervating the same cell (Hume and Purves, 1981; Purves and Hume, 1981).

A more detailed understanding of these effects has not been forthcoming, largely because the cell biology of competition is not fully understood. In general, there are two possibilities. First, axon terminals might compete with one another directly, each terminal warring with its neighbors in an effort to secure its relationship with the target cell. According to one such scheme, each nerve terminal secretes a protease capable of destroying adjacent endings (Vrbová et al., 1978). Arguing against this class of ideas, though not decisively, is the observation that the terminals from different axons are intermingled with one another on both neuronal and muscle fiber surfaces. Thus, although there is on average some separation of the presynaptic terminals from different axons on ganglion cells, they are often found side by side on the same dendrite (Hume and Purves, 1983; Forehand and Purves, 1984; Forehand, 1987). Similarly, terminals from different axons intermingle at endplate sites on muscle fibers that remain multiply innervated (Letinsky et al., 1976; Rotshenker and McMahan, 1976; Lichtman et al., 1986; Werle and Herrera, 1987). A second possibility is that axon terminals do not compete with one another directly but in a more complex way that is mediated by the postsynaptic cell. In this view, dendrites diminish competition not by simply providing more space in which terminals can get away from their neighbors, but by altering the manner in which the target cell provides trophic support to the innervation it receives. The nature of trophic support and the way

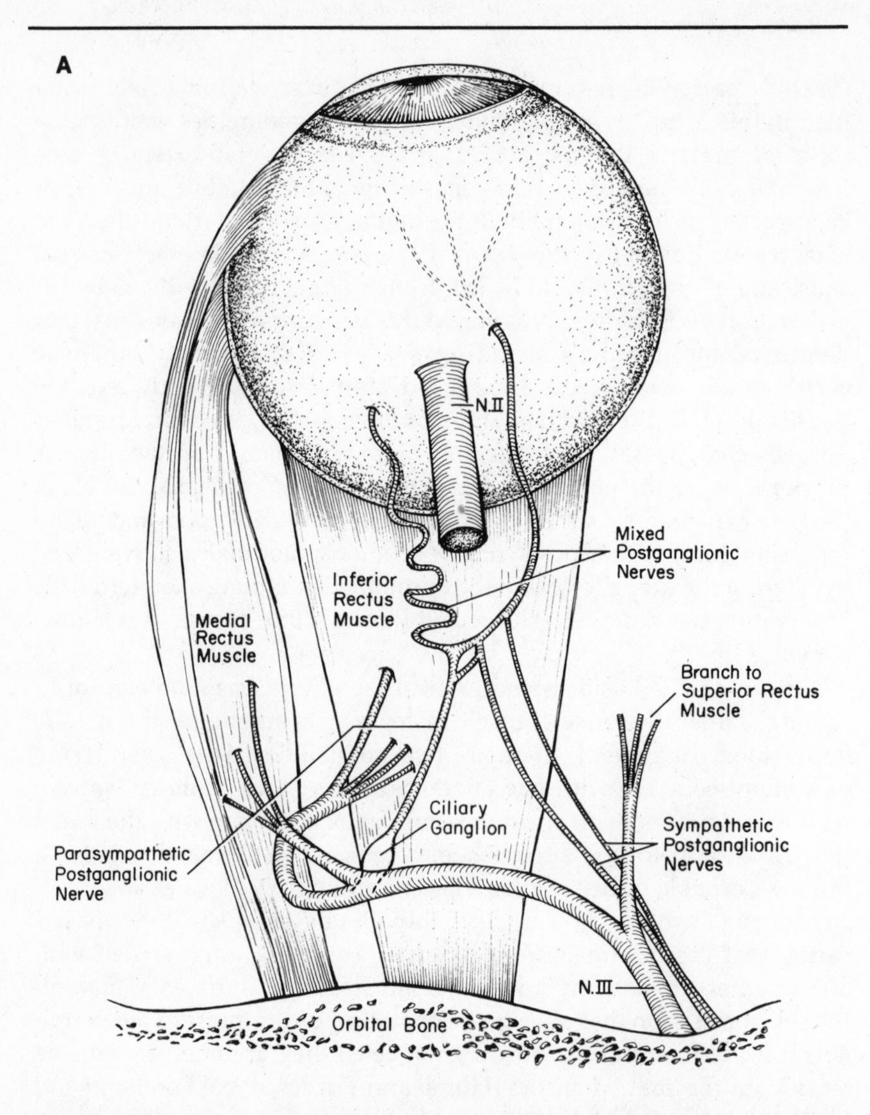

Figure 5.5. Relationship between neuronal form and input number in an autonomic ganglion. (*A*) The rabbit ciliary ganglion and its pre- and postganglionic nerves. (*B*) The number of axons innervating representative ganglion cells in adult animals. Neurons studied electrophysiologically and then labeled by intracellular injection of horseradish peroxidase have been arranged in order of increasing dendritic complexity. Comparison of the number of innervating axons (indicated by the number to the right of each ganglion cell) and neuronal shape in adult rabbits shows a strong correlation between dendritic geometry and input number. (*C*) The correlation between input number and postsynaptic geometry of the full sample, using the number of primary dendrites as an index of complexity. Each point represents the mean of measurements in a number of neurons (bars indicate the standard error); the straight line was fitted to the data by a least squares linear regression program and has a slope of about 1. (*A* from Johnson and Purves, 1981; *B* and *C* after Purves and Hume, 1981)

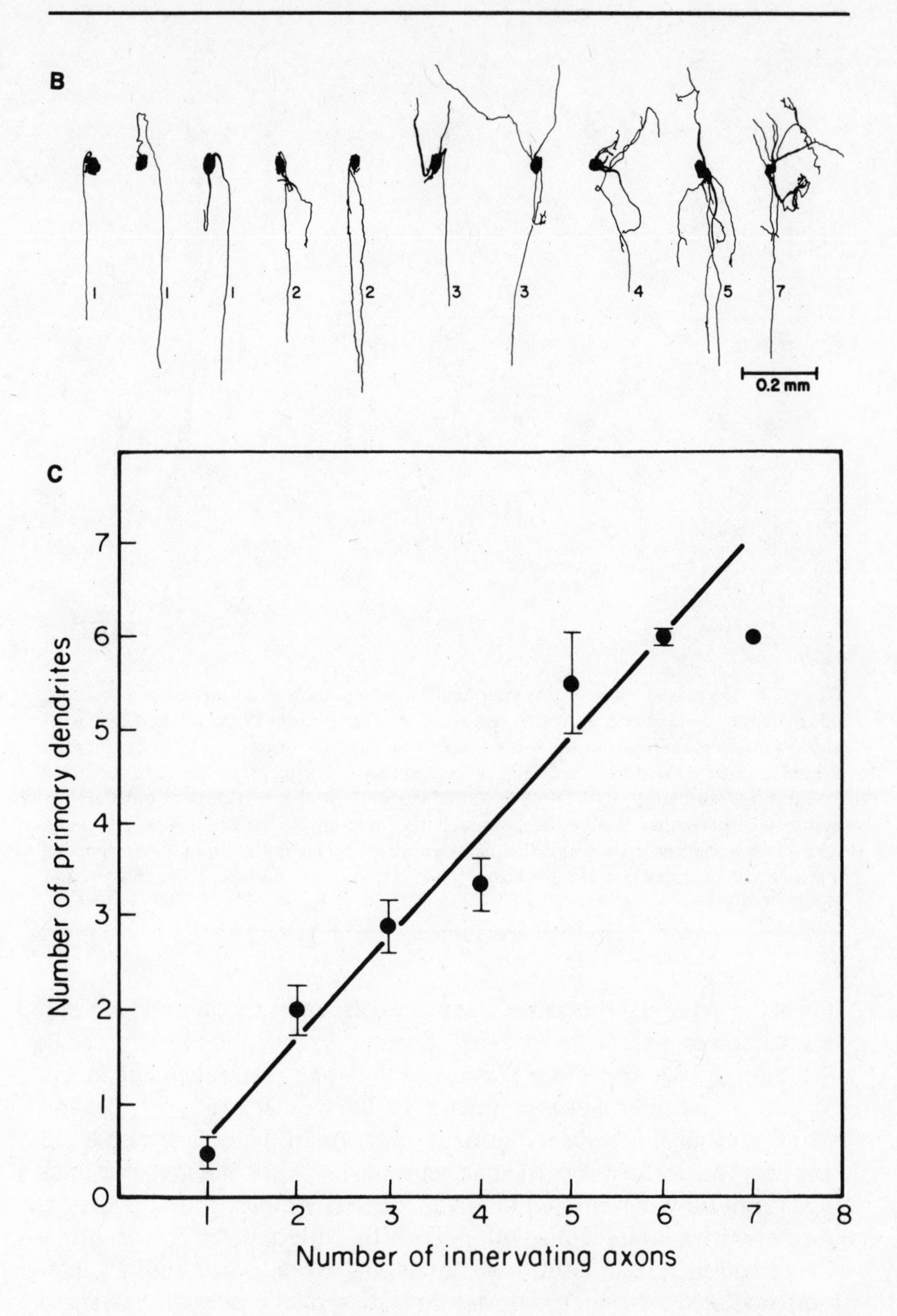
B
1
1
1
2
2
3
3
4
5
7
0.2 mm
C
Number of primary dendrites
Number of innervating axons
0
1
2
3
4
5
6
7
8

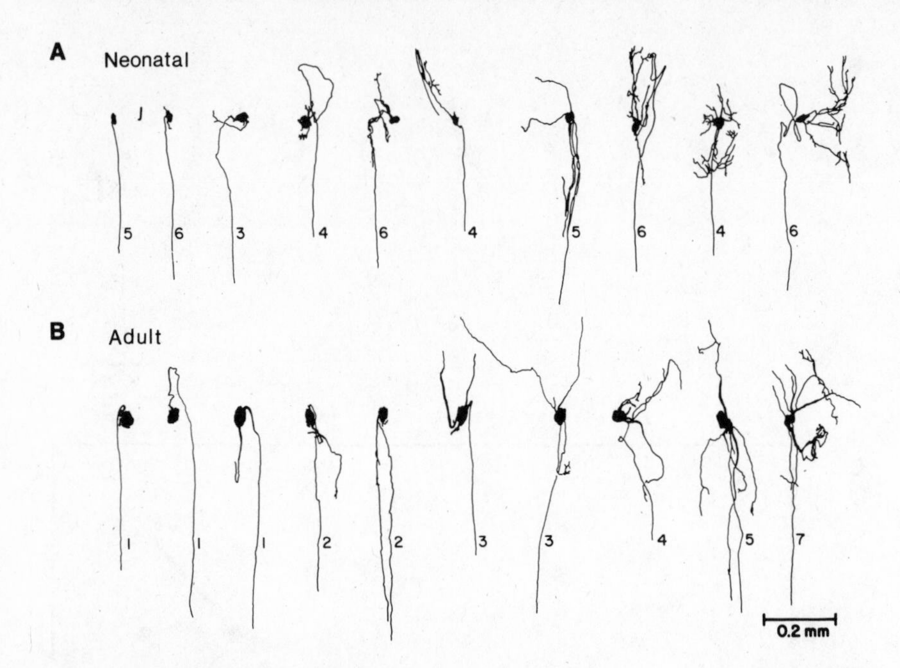

Figure 5.6. Decreased vigor of presynaptic input elimination as a function of increasing postsynaptic (dendritic) complexity. (*A*) At birth, each neuron in the rabbit ciliary ganglion is innervated by about the same number of axons (5, on average). (*B*) By the time the animal reaches maturity, a strong correlation has been established between geometry and inputs (see Figure 5.5). Comparison of the pattern of innervation at birth and in maturity implies that competitive interactions between axons innervating the same neuron are enhanced on neurons that lack dendrites and are mitigated in the presence of increasingly complex dendritic arbors. (After Hume and Purves, 1981; Purves and Hume, 1981)

it is acquired make this latter class of explanation the more likely one (see Chapters 7-8).

Although this and other aspects of synaptic rearrangement in the peripheral nervous system remain to be worked out, the general significance of these observations is clear. As in the case of nerve cell survival, the destiny of particular neuronal branches and synaptic endings is not fully determined in advance but is established in the course of competitive interactions with peers. The pattern of synaptic connections that emerges in adult vertebrates, therefore, is not simply a consequence of the biochemical identities of synaptic partners (as some early notions of chemoaffinity would have had it) or of other determinate developmental rules (which play a larger part in establishing neuronal geometry in some simple invertebrates). Rather, the adult pattern of connections, at least in the periphery, reflects a more plastic process in which neuronal connections are made and broken according to local circumstances. The broad purpose of this sort of interaction is apparently to assure the ongoing quantitative accuracy of innervation during maturation. Thus the competitive interactions that have been described in muscle and ganglia guarantee that every target cell is innervated—and continues to be innervated—by the right number of inputs and synapses, and that every innervating axon contacts the right number of target cells with an appropriate number of synaptic endings.

Synaptic Rearrangement in the Central Nervous System

The question of whether a similar rearrangement of connections takes place within the brain and spinal cord is not a straightforward one. Although numerous studies provide evidence of "plasticity" in the central nervous systems of young animals, it has rarely been possible to assess the innervation of individual central neurons, as has been done for muscle fibers and ganglion cells. The reason for this is simply the anatomical complexity of the mammalian brain; the location and arrangement of central neurons make it difficult to impale them with a microelectrode in the first place and often impossible to stimulate the full set of their inputs. In practice, most examples of plasticity in the developing central nervous system concern the establishment and maintenance of central maps rather than the innervation of individual target cells. In consequence, events in the developing brain are not easily compared to events in the periphery.

Studies of the plasticity of neural connections in the developing central nervous system fall into three categories: a small number of studies that have sought to assess inputs to individual neurons, using essentially the same approach as in the periphery; studies of the arrangement of afferent terminals within the primary visual cortex of mammals, using extracellular recording and various anatomical techniques; and studies of aberrant axonal projections in the developing brain, using anatomical tract-tracing methods. Each of these approaches has provided important information; however, the results do not have an obvious common denominator and do not lend themselves to easy summary.

With respect to the first of these categories, several studies have assessed the innervation of *individual* neurons in the developing central nervous system in order to find out whether the sort of rearrangements observed in the periphery also occur centrally. The most advantageous place to look for a reduction in the number of axons innervating a central neuron should be among cells that lack substantial numbers of dendrites and might therefore be innervated by only one or a few axons in maturity. In fact, adendritic neurons in the cochlear nucleus of the chick are subject to a process of input elimination similar to that observed among autonomic ganglion cells that lack dendrites (Jackson and Parks, 1982). However, the difficulty of impaling these neurons with a microelectrode limited the study to only a few cells. The only other class of neurons in the central nervous system that has been examined in this way is the Purkinje cell in the rat cerebellum. This neuron may seem a peculiar choice since, far from being morphologically simple, it is one of the most geometrically complex neurons found in mammals (see Figure 4.1) and is innervated by thousands of different inputs. Nevertheless, in mature mammals, each Purkinje cell is usually innervated by only a single climbing fiber (climbing fiber axons arise from the inferior olive and strongly excite Purkinje cells—Eccles et al., 1966). Consequently, it is possible to ask whether Purkinje neurons in neonatal rats are polyneuronally innervated by this particular class of inputs. As it turns out, Purkinje cells in the rat are initially innervated by several different climbing fibers which, over the first several weeks of postnatal life, are reduced to a single input (Crepel et al., 1976, 1980; Mariani and Changeux, 1980, 1981). This result is much like that observed in the muscles and autonomic ganglia of the rat.

The reduction of inputs to geometrically complex Purkinje cells is, on the face of it, at odds with observations of autonomic ganglia in which a net input elimination is all but abolished by the presence of a modest number of dendrites (see Figure 5.6). The solution of this paradox may have to do with the domain within which climbing fiber inputs compete during early life. Whereas some simple neurons, like autonomic ganglion cells (and muscle fibers), are innervated by a more or less homogeneous population of presynaptic axons, many other classes of nerve cells (Purkinje cells certainly among them) are innervated by diverse inputs that often contact only a limited region of the postsynaptic cell. Thus, in maturity, the synapses made by climbing fibers are predominantly on the shafts of Purkinje cell dendrites; at birth, however, the majority of climbing fiber synapses are found on the somata of Purkinje cells (Altman, 1972; Palay and Chan-Palay, 1974; Landis, 1988). This fact, coupled with the rudimentary nature of the dendrites arising from Purkinje cells in the neonatal rat (see Figure 4.5), may mean that a Purkinje cell does not appear all that different to a climbing fiber during the first two weeks of a rat's life than a ganglion cell without dendrites appears to its inputs. It is also possible, however, that the reduction of climbing fiber inputs to Purkinje cells is based on mechanisms that are quite different from the mechanisms of input elimination in the periphery.

The second category of studies that has addressed the issue of synaptic rearrangement in the developing brain concerns the segregation of neural connections in the visual system of young mammals. This phenomenon is best understood in the visual cortex of cats and monkeys, but it is evident in many regions of the developing brain (Hubel et al., 1977; Rakic, 1977; Shatz, 1983; Schwob and Price, 1984; Sretavan and Shatz, 1986). The most important work on the segregation of central connections in early life has been a series of studies carried out by D. H. Hubel and T. N. Wiesel on the primary visual cortex of mammals (Hubel and Wiesel, 1963, 1965; Wiesel and Hubel, 1963a,b, 1965; Hubel et al., 1977; LeVay et al., 1980; Hubel, 1982; Wiesel, 1982). This work first called attention to the malleability of neural connections at a time when the prevailing opinion was that the connections between nerve cells might be quite rigid. In adult cats and monkeys, afferent projections from the visual relay in the midbrain (the lateral geniculate nucleus) to the primary visual cortex are segregated into right and left eye regions when viewed from the cortical surface; these right and left eye stripes are generally referred to as ocular dominance columns (Figure 5.7). Although the functional significance of partitioning cortical inputs from the two eyes remains largely a mystery, the presence of ocular dominance columns in the adults of these species provides an opportunity to examine the arrangement of cortical innervation as a function of age. In both the visual cortex and lateral geniculate, axon terminals related to each eye overlap widely in late embryonic and early postnatal life (Hubel et al., 1977; Rakic, 1977; LeVay et al., 1980). As development proceeds, a gradual segregation of axons that are related to the right and left eyes can be demonstrated by either electrophysiological or anatomical means.

In several respects, this gradual rearrangement of afferents observed in the developing cortex is similar to the rearrangements described in the periphery. Both phenomena occur early in life, involve the making and breaking of synaptic connections, and appear to depend on competition. Thus, if one eye is removed at birth (or if its activation is

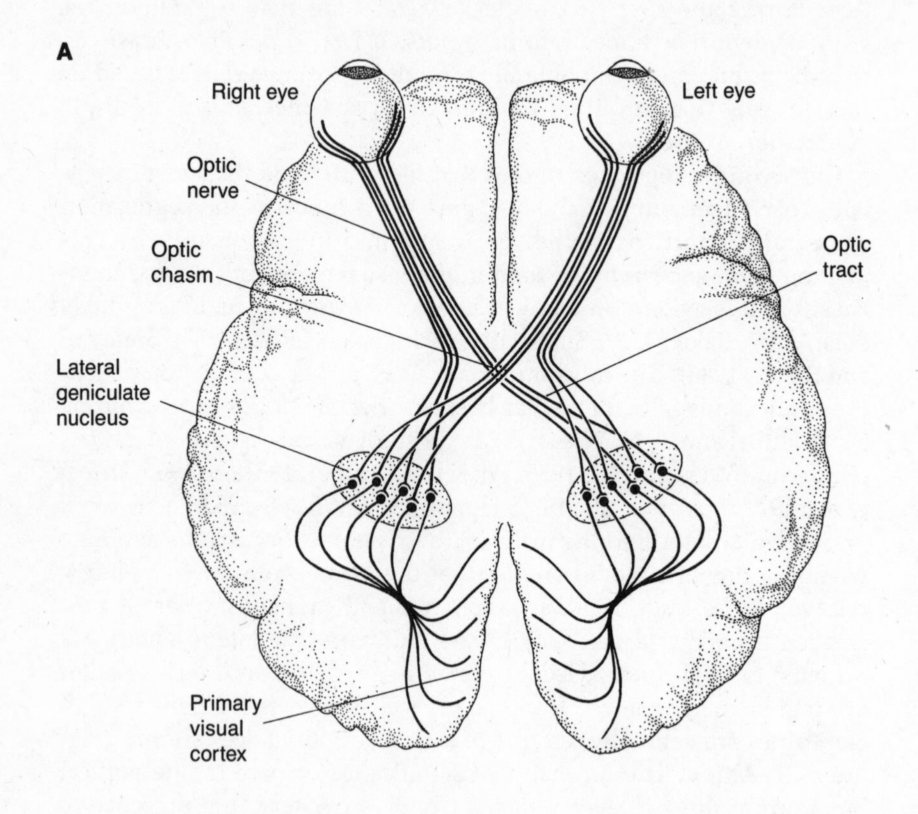

Figure 5.7. Rearrangement of afferent projections in the primary visual cortex of monkeys. (*A*) The visual pathway from the retina to the primary visual cortex in primates (seen from the ventral aspect of the brain). (*B*) Image of the cortical terminals related to an eye into which radioactive amino acids were injected. This label is incorporated into proteins which are ultimately transported to axon terminals in the primary visual cortex. In newborn monkeys, terminals related to the injected eye (white area) are widely distributed; thus, terminals from the two eyes are extensively intermingled at birth. If, however, the monkey is allowed to mature for 6 weeks or more before the labeling procedure, the slight tendency to striping seen at birth has progressed to a nearly complete segregation of inputs from the two eyes. This change in the pattern of afferent terminations indicates extensive rearrangement of neuronal branches (and presumably synapses) in this part of the maturing brain. (*C*) Reconstruction of the ocular dominance pattern over nearly the full extent of the primary visual cortex in an adult monkey. (*A* and *C* after Hubel and Wiesel, 1979; *B* after Hubel et al., 1977)

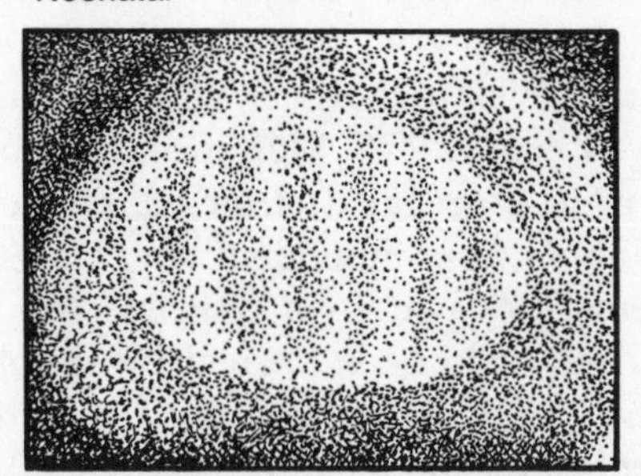
B
Neonatal

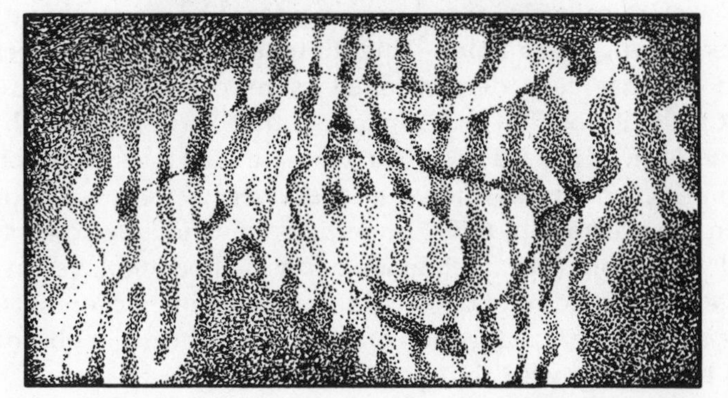
Adult
1mm

C

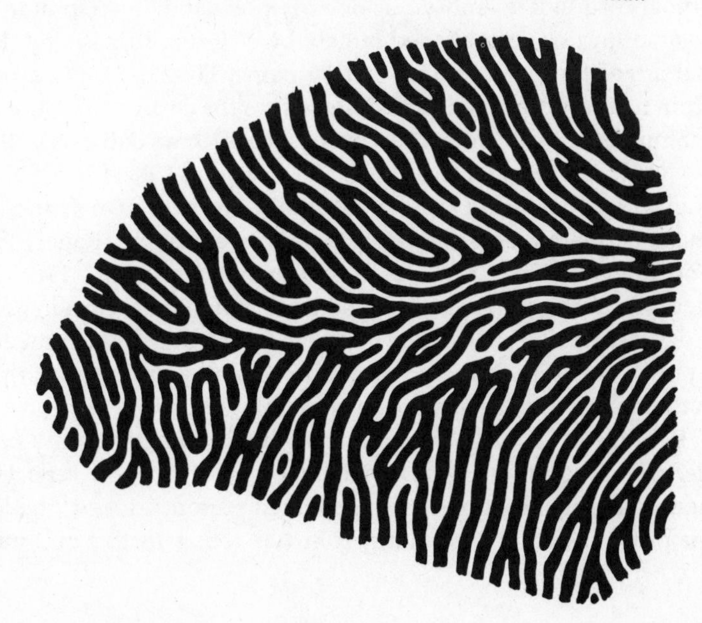

disturbed by keeping the eye closed during early life), then the normal segregation of afferents in the primary visual cortex fails to occur; the cortical projections related to the normal eye remain widely distributed, in a pattern similar to that seen at birth (Hubel et al., 1977; LeVay et al., 1978; LeVay et al., 1980). Moreover, the transition from overlapping innervation to segregated stripes almost certainly involves the elimination of inputs from individual target cells. Yet the segregation of visual afferents into patterned domains comprising thousands of target cells is unlikely to be explained by simple extrapolation from synaptic rearrangement in the peripheral nervous system (however important a part trophic interactions may play in afferent segregation).

The third category of rearrangements in the developing central nervous system of mammals—the disappearance of early aberrant projections—has been demonstrated by retrograde anatomical markers that make it possible to study axonal trajectories. Although this technique is in some ways similar to the method used to study ocular dominance columns, the issue addressed by retrograde tracing has been the accuracy of initial projections to various regions of the brain and spinal cord (the anterograde tracer method employed in the visual system was used to understand the detailed interaction of axon terminals *within* a particular region of the cortex). The major observation that has been made with retrograde tract-tracing techniques is that many central projections have a wider distribution within the brain and spinal cord in late embryonic or early postnatal life than in maturity. A related finding, established largely by electron microscopy, is that initial projections also comprise more axons. Thus, far more axons extend from one hemisphere to the other across the corpus callosum of several mammals at birth than a few weeks later (Innocenti et al., 1977, 1986; Ivy et al., 1979; Innocenti, 1981, 1982; Ivy and Killackey, 1981; Reh and Kalil, 1982; LaMantia and Rakic, 1984). Similarly, two or three times as many axons project from the retina to the midbrain (Rager, 1983; Rakic and Riley, 1983b; Crespo et al., 1985; Williams et al., 1986) and from neurons in the spinal cord and sensory ganglia to the brain (Chung and Coggeshall, 1987) at birth as in maturity. An excessive number of axons in neonatal animals is also found in some regions of the peripheral autonomic nervous system (Aguayo et al., 1973).

Because neuronal death in development occurs in many parts of the nervous system, some of this axonal excess probably reflects ongoing neuronal degeneration in the postnatal period. Indeed, the loss of the majority of axons extending from the retina to the midbrain in the embryonic or newborn rat and monkey *is* concurrent with massive death of retinal ganglion cells (Perry et al., 1983; Rakic and Riley, 1983a). However, changes in early central connections cannot be explained simply as a manifestation of neuronal death. In many instances, central projections initially extend to regions in addition to those normally innervated by the same pathway in maturity (O'Leary, 1987). Such aberrant projections may be to the wrong side of the brain, the wrong level of the neural axis, or simply the wrong part of a normal target. A good example of this misdirection occurs in the developing rat (Figure 5.8). In adult animals, no neurons in the occipital cortex send axons into the spinal cord (although many neurons from more rostral cortical regions do so). However, when a retrograde tracer is injected into the tract leading to the spinal cord at birth, large numbers of neurons in layer V of the occipital cortex are retrogradely labeled (Stanfield and O'Leary, 1985; O'Leary and Stanfield, 1986). Furthermore, because the retrograde markers persist in neurons for many weeks, they also show that some of the neurons which were initially labeled do not die (Land and Lund, 1979; Innocenti, 1981; Ivy and Killackey, 1981, 1982; O'Leary et al., 1983). In such cases, then, the "correction" of aberrant projections occurs by the loss of some axonal branches rather than by the demise of the parent cells.

The interpretation of these exuberant early projections in relation to events in the peripheral nervous system (or the visual cortex) is uncertain. Some of the phenomena demonstrated by tract tracing are apparently a result of normal cell death in the central nervous system, with death in some instances occurring preferentially among a misdirected population (O'Leary et al., 1986). In other instances, however, the elimination of axon branches clearly occurs without the death of the parent neuron. In principle, this phenomenon could be similar to the target-dependent regulation of axonal branching in the periphery. Some investigators have encouraged this interpretation by emphasizing the fact that axonal projections in the developing central nervous system are subject to a form of competition; thus, erroneous projections can be made to persist if a related projection is removed (Lund et al., 1973; Schneider, 1973; Rakic, 1981; Rakic and Riley, 1983b). However, whether exuberant (or aberrant) projections come into contact with target cells is not generally known. On that account, whether these adjustments of projections involve trophic interactions with target cells of the sort that occurs in simple peripheral systems remains an open question.

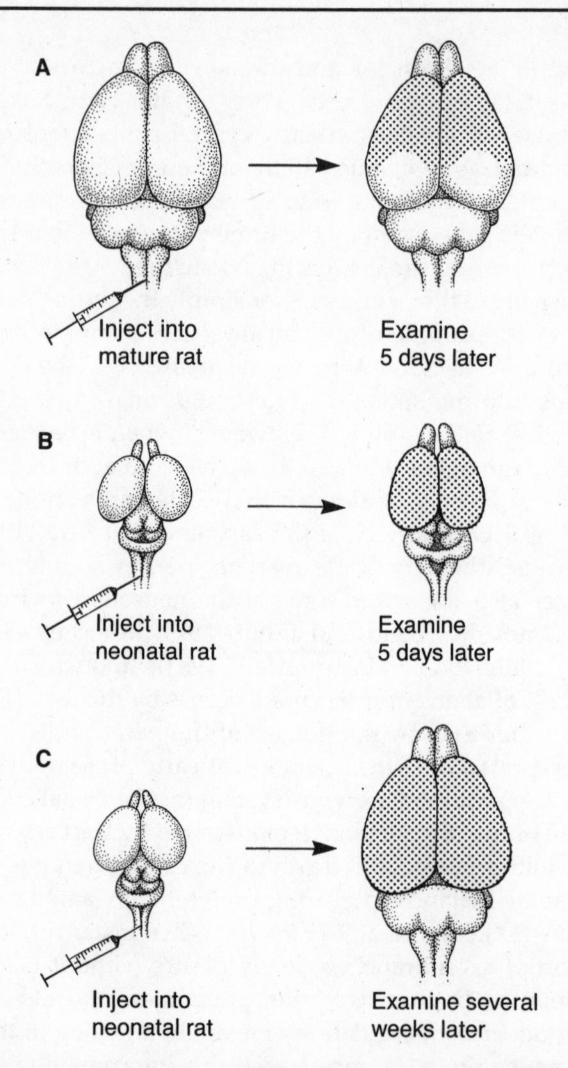

Figure 5.8. Aberrant axonal projections within the central nervous system of neonatal mammals, demonstrated with a retrograde tracing method in a rat. A dye (True blue) is injected into the corticospinal tract in the brainstem, a major pathway that carries axons from the brain to the spinal cord. (*A*) When the dye is injected into animals several weeks of age or older, the cortical neurons labeled by dye uptake over several days are restricted to the anterior two-thirds of the cerebral hemispheres. (*B*) When the same procedure is carried out at birth, however, retrogradely labeled neurons are found throughout the hemispheres. (C) That this rearrangement of the corticospinal projection is not due to cell death is shown by allowing more time to elapse between neonatal injection and examination of the brain. The presence of numerous labeled neurons throughout the cortex at an age when the inappropriate projection is no longer apparent (see A) indicates that the branches of occipital neurons extending to the spinal cord are withdrawn without degeneration of the parent neurons. (After O'Leary, 1987)

Purposes of Synaptic Rearrangement

These various rearrangements of connections (or axonal projections) serve diverse purposes. The point of the rearrangement of some initial inputs that occurs in muscles and ganglia is evidently to establish *quantitatively* appropriate innervation of *individual* target cells and, by extension, of the target as a whole. This process ensures that every axon innervates an appropriate number of target cells and that every target cell is innervated by an appropriate number of axons (that is, ensures appropriate degrees of divergence and convergence). Because the establishment of quantitatively appropriate innervation is fundamental to the operation of *any* neural circuit, this sort of regulation would be expected throughout the nervous system (just as one would expect the matching of nerve and target cell populations through neuronal death to be widespread). However, few examples of such quantitative adjustment of neural connections have actually been described in the central nervous system. The reason for this lacuna is primarily the complexity of central connections, which makes quantitative assessment of the innervation of individual cells difficult. Moreover, since most central neurons are geometrically complex, synaptic rearrangement is unlikely to be apparent by simply counting input number. As a consequence of these and other difficulties, target-dependent regulation of neuronal connections in the central nervous system remains largely inferential.

Some caution must therefore be exercised when generalizing about the rearrangement of connections in the developing nervous system of mammals or other vertebrates. Neuronal connections in the periphery appear to be governed by competitive interactions at the level of the targets that depend on trophic feedback from the target cells. It seems likely that similar trophic interactions govern neuronal connections in the central nervous system, just as the trophic basis of neuronal death in development is thought to be general. Nonetheless, one must bear in mind that information about trophic effects on neural connections is at present derived primarily from studies in the peripheral nervous system. Even if rearrangements of connections in the central nervous system are shown to depend upon trophic mechanisms, they are clearly more complex than events in the periphery. In the simplest cases, the depletion of an early axonal projection during maturation reflects neuronal death and can be considered an epiphenomenon of the matching of neuronal populations and target cell populations. In other instances, however, entire axonal projections are eliminated in early life without the death of the parent nerve cells. It is difficult to know how to regard these redundant projections that go to the wrong place. One possibility is that they represent vestigial avenues of development that were useful in a former time (in much the same way that the transient appearance of "gill clefts" and a tail in a human embryo appear to reflect phyletic history rather than important developmental strategies). Rearrangements that involve the segregation of inputs among large populations of target cells must be given yet another interpretation. Although the functional significance of segregation is for the most part unclear, its purpose is apparently different from the matching of pre- and postsynaptic populations, the quantitative adjustment of innervation, and the elimination of projection errors. The only truly common feature of these central rearrangements is that each contributes, in some way, to the accurate representation of the body within the nervous system.

Finally, the malleability of the developing vertebrate nervous system must be placed in proper perspective. Important though the phenomenon may be, the ability of neural connections to adjust to local circumstances is nonetheless a refinement superimposed on those largely determinate processes that generate the nervous system in the first place.

SIX

Regulation of Neural Connections in Maturity

THERE are obvious reasons during early development for the modification of neural circuits by interactions between neurons and the cells they innervate. Developing target cells must eventually be innervated by a functionally appropriate number of inputs and synapses, and developing neurons must eventually innervate a functionally appropriate number of target cells. At a more general level, the nervous system must establish and maintain accurate maps of the body it innervates. The rapidly changing size and form of developing animals demand that the innervation of targets, as well as the motor and sensory maps that reflect the arrangement of targets, continue to adjust throughout ontogeny. Whether this plasticity of neural connections and the mechanisms that underlie it simply cease to operate when—by some criterion—an animal is judged to have matured is a question that has not received much explicit attention. By several measures, various rearrangements of neuronal branches and synaptic connections appear to be complete shortly after birth. However, the assays on which this impression is based are relatively crude. Thus, the early stabilization of synapse counts in electron microscopical sections, of the number of axons innervating individual target cells, or of the arrangement of cortical projections does not rule out the persistence of a wide range of more subtle ongoing changes in connectivity. The resolution of this issue is of some consequence, since the perceived answer will determine how people investigate a variety of neural phenomena (in much the same way that the perception of the nervous system as being largely hard-wired influenced the investigation of neural specificity in the 1950s and 1960s). Because the evidence accumulated in recent years has so clearly demonstrated the malleability of neural connections in the early life of mammals and other vertebrates, the possibility of ongoing changes of connections in maturity has been addressed with increasing interest.

One approach to this issue has been to test whether the mature nervous system still has some *potential* for connectional changes. Accordingly, the stability of axonal and dendritic branches and their synaptic connections has been examined in the face of various experimental perturbations. These studies not only have confirmed the persistent ability of the mature nervous system to adjust to an experimental challenge but also have given insight into the cellular mechanisms of trophic interaction. A major point to emerge from the perturbation of neural connections in maturity is that patterns of connection represent a *balance* of the various forces involved in the regulation of neuronal branching and synapse formation. This work has also led to efforts to determine, by direct observation, whether this balance is static or dynamic.

Target-Dependent Neuronal Survival

During development, the most obvious expression of trophic dependence is the survival or death of neurons as a function of interactions with their targets; thus, if neural targets are removed or augmented in early life, the number of surviving neurons changes accordingly. Although it is generally accepted that the number of neurons is more or less fixed in the nervous system of mature vertebrates under normal circumstances, it is possible to test for persistent target dependence in maturity by assessing neuronal survival after removing a target or otherwise interrupting the link between neurons and the cells they innervate.

Most observations relevant to this issue have been made in the mammalian motor system. Anatomists at the turn of the century were aware that limb amputation in man causes a depletion of the related spinal motor neurons and that some cortical neurons synaptically linked to the spinal motor neurons are also lost after amputation (Campbell, 1905). More detailed experimental studies in animals, however, have made clear that limb amputation or motor nerve section has less severe consequences in mature animals than in neonates. Thus, amputation or nerve section causes prompt and extensive neuronal loss in newborn mammals but usually only limited and gradual loss in mature animals (Romanes, 1946; Schmalbruch, 1984; Kashihara et al., 1987). Studies of motor neurons in the mammalian autonomic system have yielded comparable results. If a postganglionic nerve is cut early in life, the majority of affected ganglion cells die promptly (Hendry, 1975; Hendry and Campbell, 1976). If, however, the same postganglionic nerve is interrupted in maturity, only about half the ganglion cells die, and then only over a period of several months (Purves, 1975). In the guinea pig vagal nucleus, the loss of visceral motor neurons after interruption of the vagus nerve is gradually progressive over a year or more (Laiwand et al., 1987). And in man, the full effect of limb amputation on the numbers of related motor neurons in the spinal cord is evident only after many years (Kawamura and Dyck, 1981).

Assessment of neurons that project entirely within the brain of mammals also indicates ongoing target dependence in maturity (Cowan, 1970). If anything, the deleterious effects of interrupting axons in central pathways appear to be greater than those of interrupting axons that project peripherally. For example, at least half the cholinergic neurons of the basal forebrain degenerate within a month of cutting their projections to the hippocampus (Hefti, 1986; see also Goshgarian et al., 1983). A large number of retinal ganglion cells also die in adult rats after optic nerve section (Aguayo et al., 1987). The gist of these observations is that many mammalian neurons continue to depend on their targets for survival in maturity, even though the intensity of this dependence often diminishes with age.

Experimentally Induced Changes of Connectivity

Axonal and dendritic processes are also subject to ongoing trophic regulation in maturity. Evidence for the continued effect of targets on neuronal branches and synapses derives in large part from manipulations that stimulate axons and dendrites to renewed growth and the formation of novel connections. Complementary evidence comes from manipulations that cause a withdrawal of branches and synaptic connections and from indications that these two processes (growth and retraction) are related in a reciprocal manner. In addition to showing that the mature nervous system retains the ability to make and break synaptic connections, these observations imply that patterns of innervation are normally in a state of balance. In this sense, patterns of neural connections in maturity are actively *maintained*.

A variety of stimuli induce mature axons to extend their terminal arborizations and to form new synaptic contacts. The best studied examples of this process are in the peripheral nervous system. Evidence for sprouting is widespread, however, and this ability is probably a general property of mature vertebrate neurons (but see Rodin et al., 1983). The most easily interpreted manipulation that induces axonal sprouting in maturity is partial denervation (Figure 6.1). If an

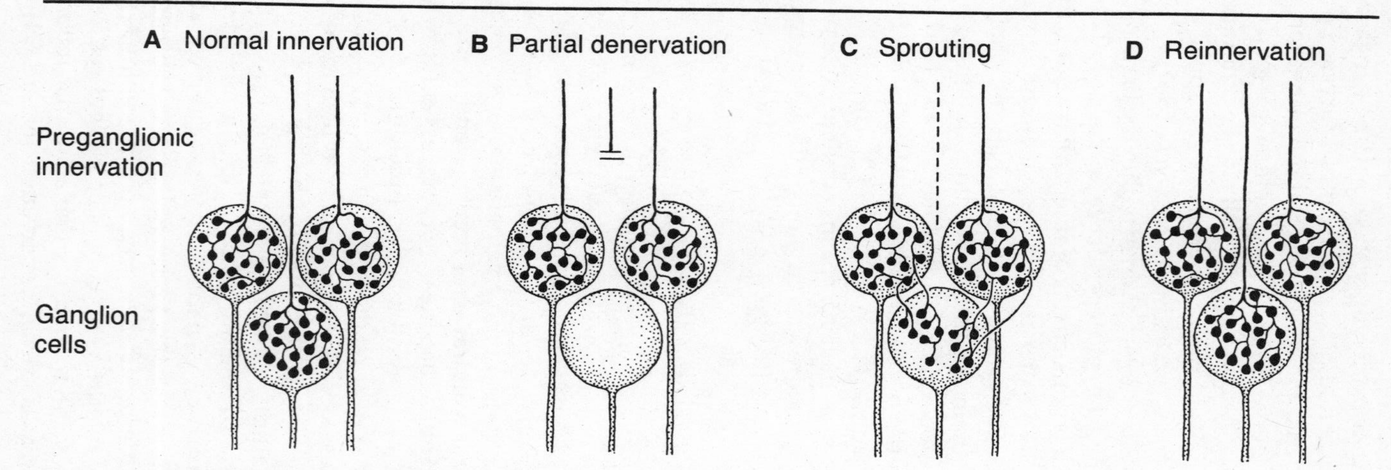

Figure 6.1. Renewed terminal growth and synaptogenesis among residual axons following partial denervation of target cells. (*A*) The normal arrangement of innervation among adendritic ganglion cells. (*B*) When a fraction of the preganglionic axons is cut, some neurons are denervated. (*C*) Within a few days, the residual (intact) axons send out new terminal branches which innervate the denervated neurons. The fact that the residual terminals are stimulated at a distance suggests a diffusible signal arising from the denervated cells. (*D*) When the interrupted axons regenerate, the sprouts give way in favor of the original axons. This phenomenon presumably represents correction of a quantitative imbalance which occurs when axons with relatively few terminals compete with axons that have made a surfeit of connections. The entire process is in some ways similar to the initial innervation and rearrangement of connections that occurs in normal development.

innervated target is deprived of some but not all of its innervation, the residual axon terminals respond, over a period of a few days, by proliferating new terminal branches and in many instances by forming new synaptic connections (Edds, 1953; Murray and Thompson, 1957; Courtney and Roper, 1976; Roper and Ko, 1978; Brown et al., 1981b; Maehlen and Njå, 1981; Maehlen and Njå, 1984; see also Cotman et al., 1981; Mendell, 1984, for reviews of sprouting in the central nervous system). Indeed, if an autonomic ganglion is completely denervated, the ganglionic neurons may form synapses with one another (Sargent and Dennis, 1977; Johnson, 1988).

Thus, mature neurons certainly have the capacity to form additional synapses. If, however, the original innervation is allowed to return, the overextended axon terminals lose their extra synapses and tend to re-establish their original field of innervation (Guth and Bernstein, 1961; Roper and Ko, 1978; Wigston,1980; Maehlen and Njå, 1981, 1984; Sargent and Dennis, 1981; Gorio et al., 1983). This redress of the synaptic imbalance elicited by partial denervation implies an equilibrium, in normal circumstances, between those influences that stimulate axon growth and synaptogenesis and those influences that favor an involution of terminal branches and synaptic loss. These effects have generally been interpreted in terms of a disruption of the normal trophic signaling that goes on between pre- and postsynaptic elements during the establishment and maintenance of synapses (Harris, 1974; Purves, 1977; Purves and Lichtman, 1978, 1985b; Brown et al., 1981b; Brown, 1984; see also Diamond et al., 1976; Thompson, 1985).

Experimental intervention can also cause the withdrawal of a portion of a cell's *normal* complement of synaptic connections in maturity. An example of such involution comes from experiments in which the connections between neurons and their targets are interrupted. Disconnecting a neuron from its target results in a loss of terminals from the parent nerve cells. This conclusion has been established in a number of neural systems, notably on spinal and brainstem motor neurons that project to skeletal muscles (Downman et al., 1953; Eccles et al., 1958; Kuno and Llinás, 1970; Sumner and Sutherland, 1973; Watson, 1974; Sumner, 1975, 1976; Mendell et al., 1976; Faber, 1984) and on autonomic neurons whose axons project to visceral targets (Matthews and Nelson, 1975; Purves, 1975, 1976a; Brenner and Johnson, 1976; but see Gordon et al., 1987).

To take autonomic ganglia as the simpler example, cutting or otherwise interrupting the postganglionic nerves of sympathetic ganglia causes a rapid and profound loss of synapses from the surfaces of the ganglion cells (Figure 6.2; Matthews and Nelson, 1975; Purves, 1975). Within a week after axotomy, as these procedures are collectively called, about two-thirds of the synapses made by preganglionic axons on the injured neurons are lost. A similar loss of at least one class of afferents from spinal motor neurons is evident when peripheral motor nerves are cut (Mendell, 1984). It seems unlikely that such a loss represents a response to injury *per se,* that is, to the presumptive "ill-health" of neurons coping with the task of regenerating peripheral processes (Lieberman, 1971; Brodal, 1982). The reason for this assumption is that synapse loss after axotomy can be mimicked by local application of drugs such as colchicine which interfere with axoplasmic transport in the postganglionic nerves (Pilar and Landmesser, 1972; Purves, 1976a). The rationale of these latter experiments is to abrogate communication between nerve cells and targets without mechanical interruption, interference with electrical signaling, or requiring the nerve cell to reorganize its synthetic machinery to elaborate a new axon (as occurs after surgical interruption of peripheral nerves).

The synaptic loss that occurs after disrupting the link between nerve cells and their targets is reversible (Figure 6.2). Thus, when autonomic axons regenerate (a feat they readily accomplish), the synaptic connections on the injured neurons are almost completely restored. If by some permanent means, however, ganglion cells are prevented from reaching their target (for example, a ligature left in place which keeps the majority of axons from regenerating), then over a period of several months most of these cells die (Purves, 1975). As in the case of sprouting, the loss of neuronal synapses after disruption of the integrity of the postsynaptic cell's link with its target has generally been interpreted in terms of a disturbance of the normal trophic interactions between pre- and postsynaptic neurons (Watson, 1974; Purves, 1975; Purves and Njå, 1978; Purves and Lichtman, 1985b; Eccles, 1986).

Experimentally induced changes in neural connectivity are not limited to effects on presynaptic branches: dendrites, as well as axon terminals, are subject to ongoing regulation in maturity. When skeletal or visceral motor neurons are transiently cut off from their targets, their dendrites undergo involutional changes and later sprouting (Figure 6.3; Cerf and Chacko, 1958; Grant, 1965, 1968; Grant and Westman, 1969; Sumner and Watson, 1971; Purves, 1975; Standler and Bernstein, 1982; Yawo, 1987). Thus, within a week of interrupting the connection between sympathetic ganglion cells and their targets, the dendritic arbors of many neurons begin to shrink; within about two weeks, most ganglion cells show evidence of dendritic retraction (Yawo, 1987). As

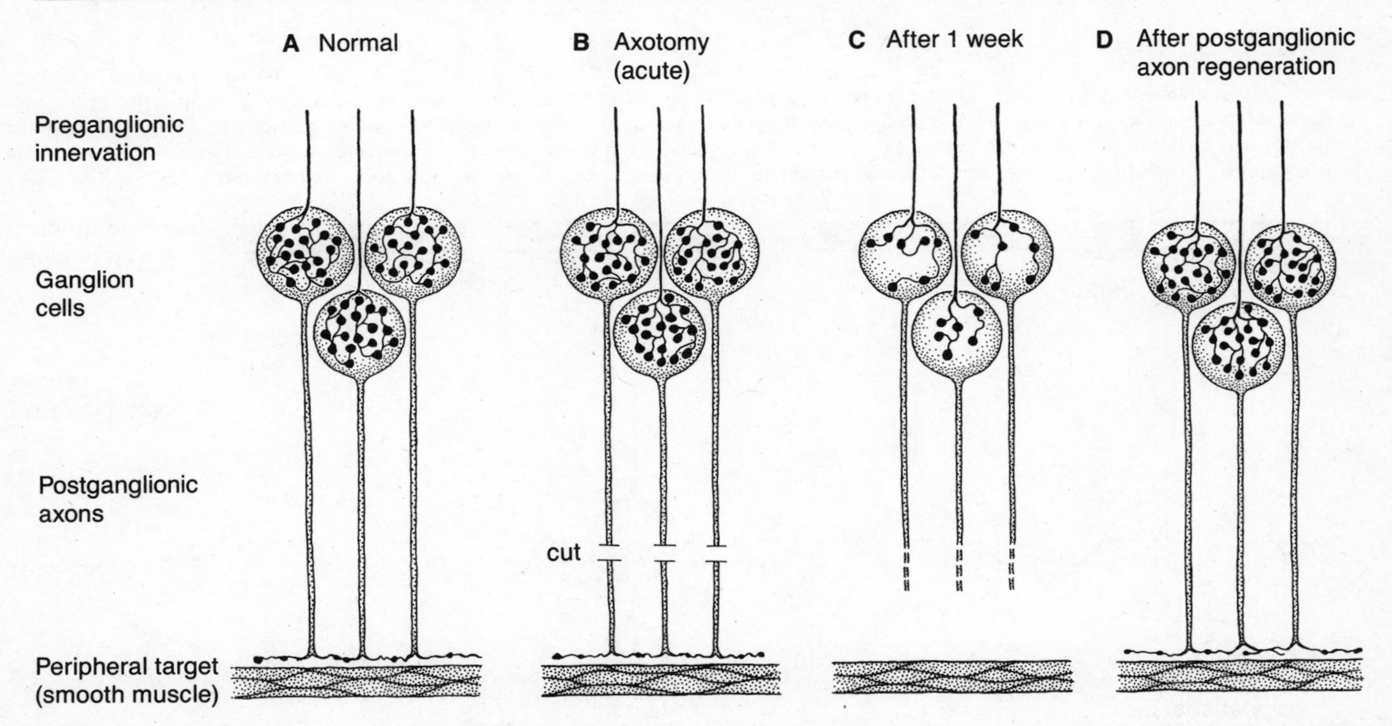

Figure 6.2. Loss of synaptic connections from ganglion cells following interruption of the link between neurons and their peripheral targets. (*A*) The postganglionic axons of ganglion cells extend peripherally to innervate smooth muscle or other visceral targets. (*B*) When the normal connection between ganglion cells and targets is interrupted, the axons distal to the lesion degenerate. (*C*) Most (but not all) of the synapses on ganglion cells are lost within a week of postganglionic axotomy. (*D*) The synaptic loss after axotomy is reversible. Thus, full recovery of synaptic function occurs over several weeks, roughly coincident with the regeneration of postganglionic axons to the periphery. This sequence of events indicates an ongoing dependence of ganglionic synapses on some quality of the postsynaptic neuron which depends in turn on the integrity of postganglionic innervation.

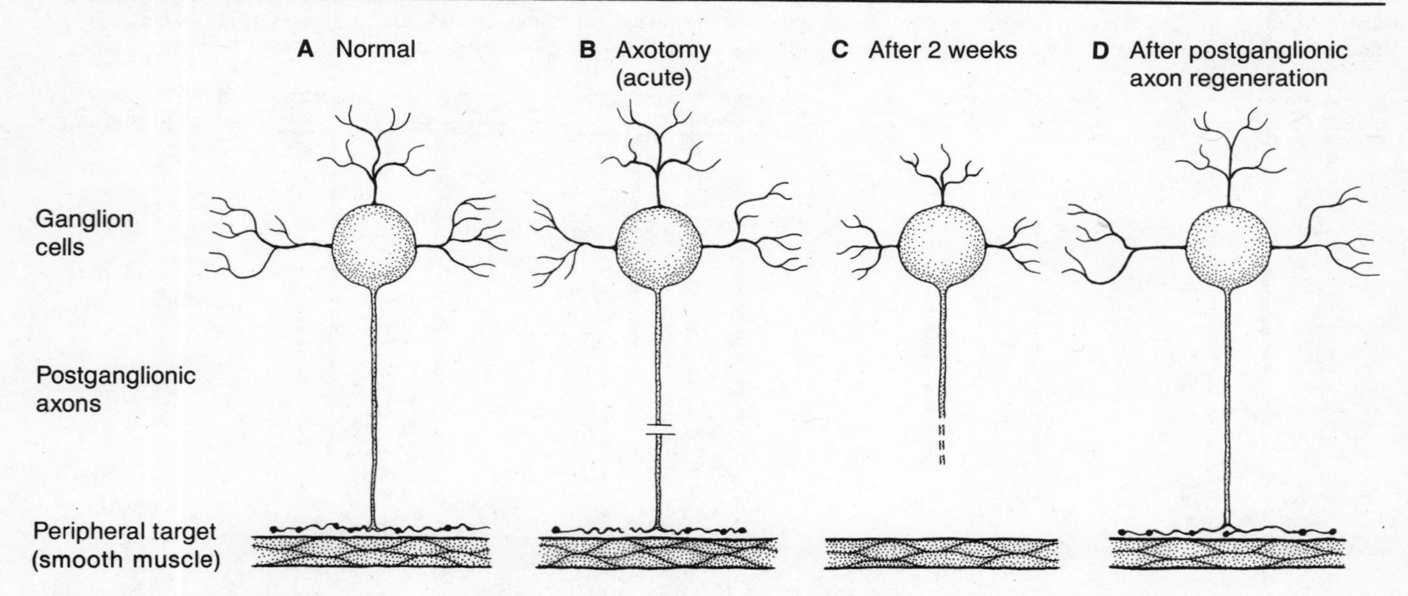

Figure 6.3. Effects of postganglionic axotomy on the dendritic morphology of sympathetic ganglion cells. (*A*) Normally, neurons in the superior cervical ganglion of mammals have substantial dendritic arbors. (*B-C*) Within two weeks of the cutting of axons that link the ganglion cells to their peripheral targets, the dendritic arbors of the neurons shrink by 60 percent, on average. (*D*) This effect is reversible. The length and complexity of ganglion cell dendrites is gradually reestablished over a period that corresponds with the regeneration of axons to peripheral targets. (Based on Yawo, 1987)

in the case of synapse loss after axotomy, the effect of postganglionic axon interruption on dendrites is reversible over a period that corresponds roughly with the regeneration of axons to the periphery (Figure 6.3C-D; see also Cerf and Chacko, 1958; Sumner and Watson, 1971; Standler and Bernstein, 1982). Parenthetically, this shrinkage of dendrites is unlikely by itself to cause preganglionic synapse loss after postganglionic axotomy. First, the changes in dendritic arbors have a slower time course than does synaptic withdrawal (Purves, 1975; Yawo, 1987). Second, postaxotomy synapse loss of the same general magnitude occurs among parasympathetic neurons that lack dendrites altogether (Brenner and Johnson, 1976).

The idea that dendritic form remains responsive to changes in the connectivity of the parent neuron is already implied by the observation that the dendritic arbors of ganglion cells continue to extend over a period and at a rate that parallel the overall growth of an animal like the rat (see Figure 4.5). This inference has been greatly strengthened by a related series of studies in which the ratio of innervating neurons to target cells is experimentally altered (Figure 6.4). The number of neurons in a sympathetic ganglion like the superior cervical can be permanently changed by taking advantage of the fact that neurons cut off from their targets in early life degenerate. Thus, partial cutting of the postganglionic nerve that innervates the salivary gland in a rat reduces the number of neurons projecting to this target to about half its normal value (Voyvodic, 1987a and unpublished). The remaining neurons (which are intermingled with ganglion cells that project to other targets) can be identified by retrograde labeling with a fluorescent dye injected into the salivary gland; the geometry of these neurons can then be assessed by intracellular injection of a dendritic marker like horseradish peroxidase into the fluorescent cells. The surviving neurons, which have a relatively greater than normal amount of target tissue available to them, and which therefore undergo axonal sprouting at the level of the target, have much larger dendritic arbors than neurons innervating a normal salivary gland. The converse experiment, in which the size of the target is reduced, can be carried out by ligating the salivary duct, thus causing the gland to atrophy. In this circumstance, a normal number of neurons is made to innervate a smaller target. Impalement and labeling of these cells in maturity shows them to have abnormally small dendritic arbors.

Experiments such as these, which reveal the ongoing ability of axonal and dendritic branches to respond to artificial imbalances of connectivity, have a significance that goes beyond the demonstration of a

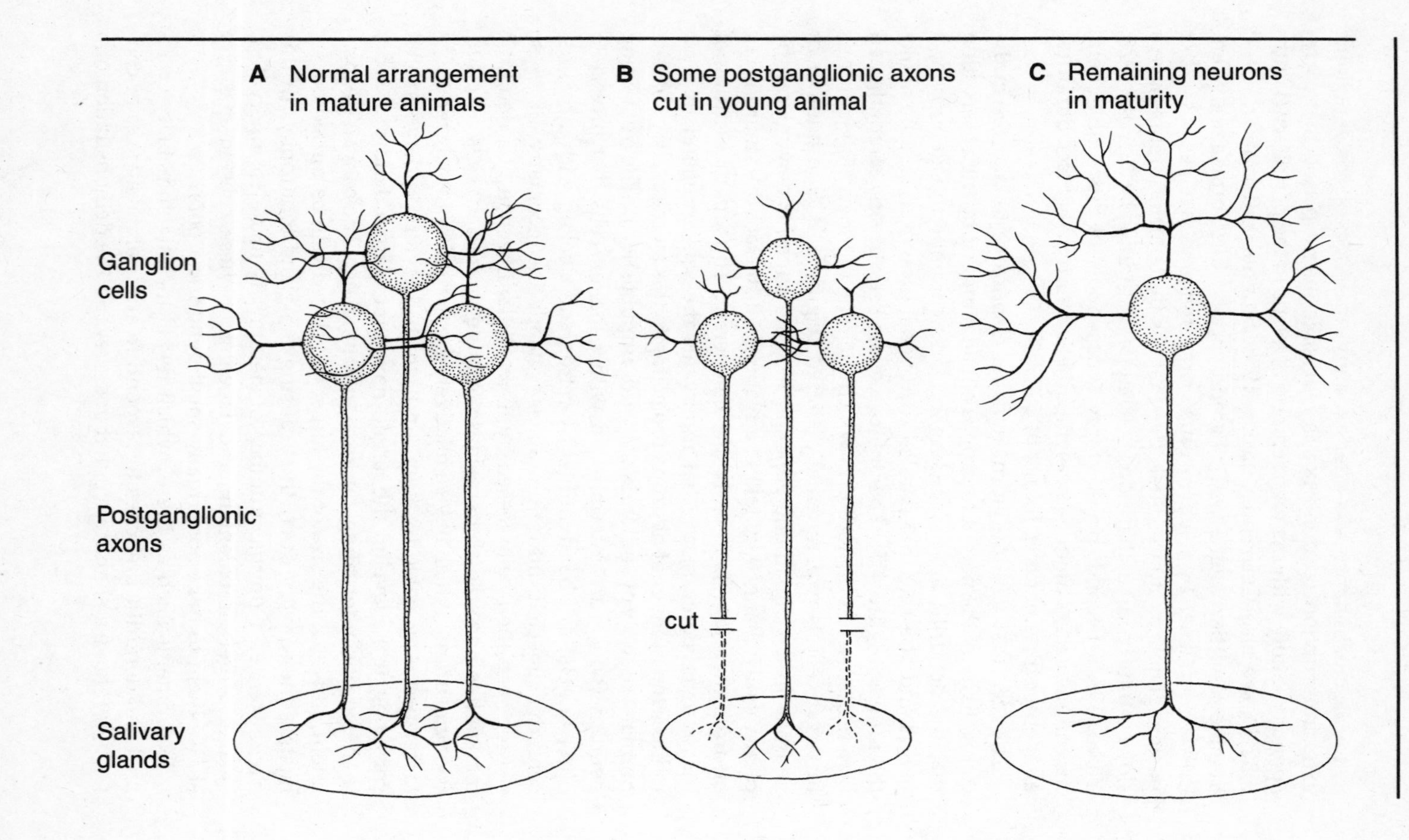
A Normal arrangement in mature animals
B Some postganglionic axons cut in young animal
C Remaining neurons in maturity
Ganglion cells
Postganglionic axons
Salivary glands
cut

potential for change in maturity. They bear out less direct arguments for growth-related adjustments of neuronal branches given in earlier chapters. A reorganization of axonal and dendritic branches really *does* occur when the ratio of neurons to target cells changes at any point in life.

Normal Remodeling of Axonal and Dendritic Branches

Evidence that synaptic connections in the vertebrate nervous system are maintained in an equilibrium that gradually changes as the size and form of animals change raises the further question of whether this equilibrium is, by nature, static or dynamic. In the absence of experimental intervention or the demands of rapid growth, synaptic connections might be quite static. Alternatively, extension and retraction of dendritic and axonal branches might occur continually in the mature nervous system. The distinction is important because it bears not only on the cellular mechanisms of connectional change but also on the kinds of uses that might normally be made of the plasticity of neural connections.

The idea that synaptic connections might be continually remodeled is not new. As long ago as the 1920s, consideration was given to the possibility that trophic interplay might alter neural connections (Ramón y Cajal, 1929). However, the notion of continual remodeling was first suggested explicitly in the 1960s in anatomical studies of the configuration of motor nerve terminals in adult mammalian muscle (Barker and Ip, 1966; Tuffery, 1971). In this work silver-stained terminals often gave evidence of what appeared to be delicate sprouts extending away from the region of the endplate. It was suggested, therefore, that these sprouts might indicate a continual process of extension and re-

---

Figure 6.4. Response of neuronal dendrites to the innervation of a larger periphery by the parent neuron. (*A*) Normally, about 15 percent of neurons in the rat superior cervical ganglion innervate the submandibular and sublingual salivary glands. (*B*) The ratio of these neurons to the number of target cells can be decreased by cutting a portion of the postganglionic axons at birth. In young animals, the neurons whose axons are cut die, presumably because they are prevented from receiving trophic support from the target. As a result, fewer neurons are available to innervate the normal target. (*C*) The remaining neurons that innervate the salivary glands have much larger dendritic arbors than normal, as shown by intracellular labeling. The innervation of a greater number of target cells by a neuron evidently stimulates a corresponding change in its dendritic configuration. (Courtesy of J. T. Voyvodic)

traction of terminal elements. This general idea was given further impetus when observations consistent with normally occurring degeneration and regeneration of nerve endings were described in visceral muscle (Townes-Anderson and Raviola, 1978; Owman, 1981) and in several regions of the adult mammalian brain (Sotelo and Palay, 1971). Finally, changes in the complexity and arrangement of neuromuscular junctions and motor units in old age imply remodeling of muscle innervation in the absence of growth or experimental perturbation (Wernig and Herrera, 1986).

In spite of the long-standing interest of neurobiologists, the issue of neuronal remodeling has been taken up in earnest only recently. So far, all of the relevant studies have been carried out in the peripheral nervous system, where the accessibility and simplicity of innervation favor a decisive analysis. Much of this work has continued to examine the innervation of skeletal muscle.

As a class, studies of muscle can be divided into those that have addressed the stability of postsynaptic elements (collectively called the endplate) and those that have addressed the stability of the motor nerve endings themselves. To take the postsynaptic aspect first, if a motor nerve to a muscle in a mammal or other vertebrate is cut, the endplate—which normally occupies only a small fraction of the fiber length—remains intact for many weeks. This fact can be demonstrated by the persistence at the original site of acetylcholinesterase, acetylcholine receptors, and other synapse-specific postsynaptic specializations such as junctional folds (Gutmann and Young, 1944; Birks et al., 1960; Letinsky et al., 1976; Steinbach, 1981; Steinbach and Bloch, 1986). Moreover, if the severed axons are allowed to regenerate, then the regrowing sprouts proceed to the original sites and reoccupy the postsynaptic specializations with a high degree of accuracy (Letinsky et al., 1976; Sanes et al., 1978; Sanes et al., 1980; Rich and Lichtman, 1986). These observations suggest that the postsynaptic structure of the mature neuromuscular junction is a relatively fixed complex which normally shifts its position little, if at all, over periods of several weeks or longer.

More detailed studies of the relationship of motor nerve endings and postsynaptic specializations, however, have supported the original contention made in the 1960s that some remodeling is a normal feature of the vertebrate neuromuscular junction. If the pre- and postsynaptic elements are conjointly stained at frog neuromuscular junctions, then regions where the nerve terminals seem to have extended beyond the endplate—or regions where a terminal appears to have withdrawn—

are often found (Figure 6.5; Mallart et al., 1980; Wernig et al., 1980a,b, 1981; Anzil et al., 1984; Bieser et al., 1984; Wernig et al., 1984; Herrera and Scott, 1985; Wernig and Herrera, 1986). In some instances, a presynaptic terminal is present but the corresponding postsynaptic specialization is missing. Such regions are taken to represent a situation in which a terminal has recently sprouted but has not yet had time to induce a corresponding postsynaptic change. In other cases, a postsynaptic specialization lacks a fully congruent presynaptic nerve terminal. These instances are interpreted as endplate sites from which a portion of the axon terminal has recently withdrawn.

These inferences about remodeling of the neuromuscular junction based on the examination of fixed material are subject to a number of uncertainties. A more direct approach to assessing neuromuscular re-

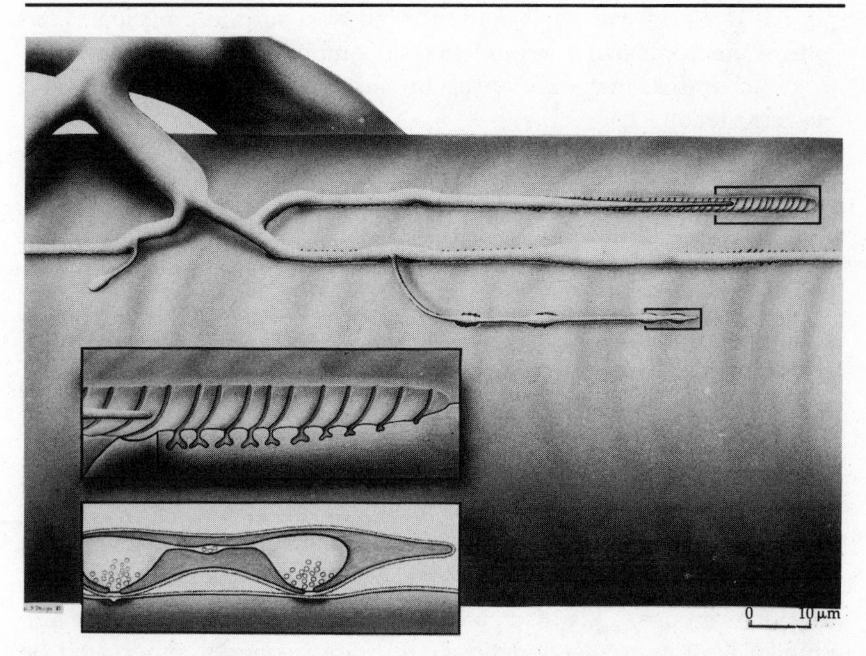

Figure 6.5. Terminal remodeling in mature amphibians, inferred from histological examination of the neuromuscular junction. The terminal branches of a motor axon innervating a single muscle fiber are shown in the frog (main figure). In some instances (small upper box), a presynaptic terminal appears to be retracting from the corresponding postsynaptic specialization. In other portions of the synapse (small lower box), a branch appears to be extending, having not yet induced corresponding specializations on the muscle fiber surface. The two boxed areas are enlarged in insets below. (From Wernig and Fischer, 1986)

modeling would be to follow individual junctions prospectively. A series of recent technical advances has made possible observations of the same identified neurons over time in both muscle and autonomic ganglia. These advances involve the use of low-light-level video microscopy, image processing, and vital dyes which can stain nerve cells and their processes without causing appreciable damage (Purves, Hadley, and Voyvodic, 1986; Voyvodic, 1986; Magrassi et al., 1987; Purves and Voyvodic, 1987). Such techniques allow the experimenter to make an image of the configuration of a particular neuron in an anesthetized animal; the procedure can then be repeated days, weeks, or months later. Examination of the frog neuromuscular junction with these methods has confirmed that remodeling normally occurs on many muscle fibers in this animal (Herrera and Banner, 1987). Thus, identified branches of the terminal arbor on about half the fibers in the cutaneous pectoris muscle were observed to extend or retract over a period of several weeks. Some uncertainty about the prevalence and rate of neuromuscular remodeling in amphibians persists because frogs show seasonal differences in this phenomenon (Wernig and Herrera, 1986).

In mammalian muscle, a less dramatic result is observed. When a large number of neuromuscular junctions were followed for periods of up to 6 months in the sternomastoid muscle of adult mice, less than 1 percent of the junctions were found to show obvious signs of remodeling (defined, as in the frog, by the establishment of novel terminal branches or the withdrawal of existing ones). Virtually all of the terminals examined, however, showed proportional growth, resulting in an increase in the overall length of the terminal branches by an average of 30 percent over 5–6 months (Figure 6.6; Lichtman et al., 1987). A greater degree of remodeling has been reported in another muscle in the mouse, the soleus, in which about 20 percent of the neuromuscular junctions show some remodeling over several months (Wigston, 1987; see also Cardasis and Padykula, 1981; Wernig et al., 1984).

It thus appears that neuromuscular remodeling is more vigorous in frog than mouse. This discrepancy may arise from the fact that amphibian neuromuscular junctions are more frequently innervated by multiple axons than are mammalian skeletal muscle fibers (Werle and Herrera, 1987) or that they are less active. In accord with the possibility that multiple innervation may influence remodeling, when endplates in the mouse sternomastoid are reinnervated by a single axon after experimental nerve injury, the regenerating axon tends to reoccupy the former endplate site quite exactly. If, however, reinnervation results in

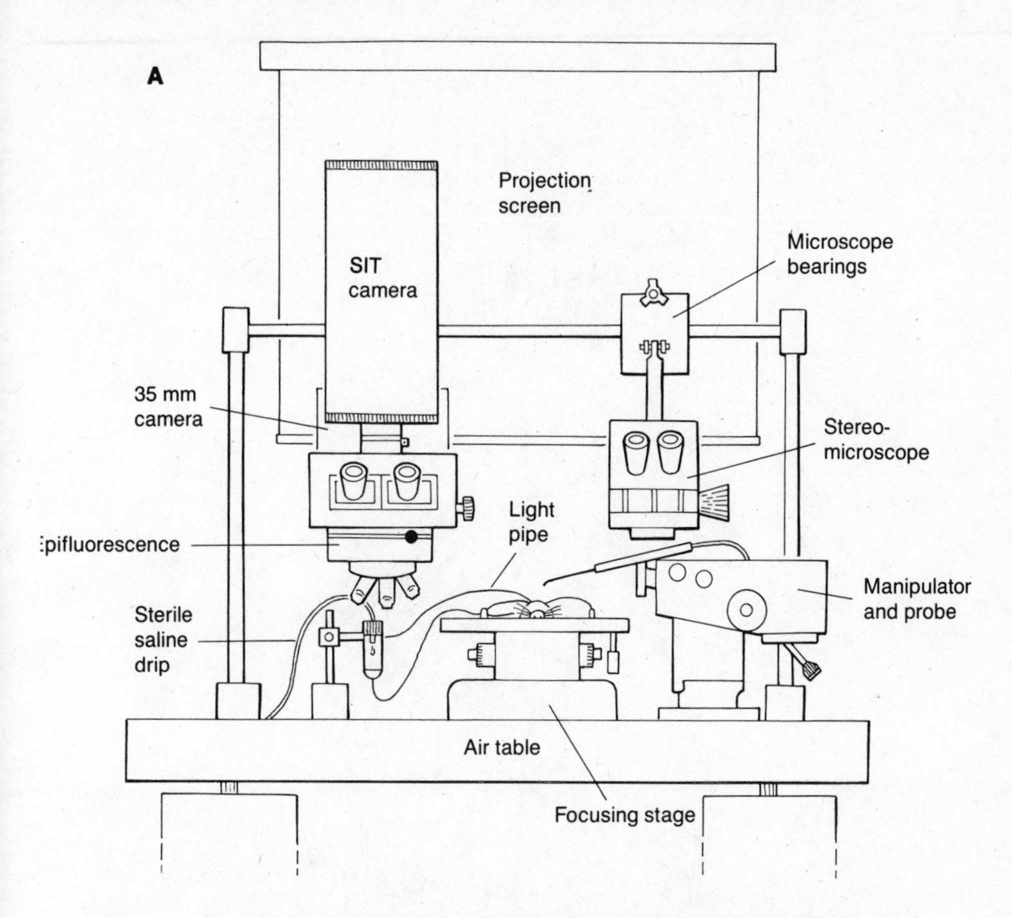

Figure 6.6. Prospective study of identified mammalian neuromuscular junctions over a period of several months in an adult mouse. (*A*) Apparatus for viewing individual axon terminals or dendrites in the peripheral nervous system of a living animal. An anesthetized mouse is placed on a stage that can be moved by micrometer drives in all three axes. To protect the living cells from photodamage, a low-light-level video (SIT) camera is used; the faint images obtained in this way are enhanced by the use of an image-processing system. (*B*) The muscle to be examined, the sternomastoid, is exposed by minor surgery and viewed with epi-illumination using a compound microscope and water immersion objectives. (*C*) The muscle is gently lifted away from the underlying tissue, and the surface is flattened slightly with a coverslip. The motor nerve terminals, located in the midportion of the muscle, are made visible by topical application of a fluorescent vital dye of the styryl pyridinium family. (*D*) The configuration of a motor nerve terminal after topical vital staining for several minutes. The overall length of the component branches of the terminal as initially observed (left) increases substantially over 5 months, from 387μm to 522μm (right). The configuration of individual branches, however, is only slightly changed. (*A* after Purves, Hadley, and Voyvodic, 1986; *B-D* from Lichtman et al., 1987)

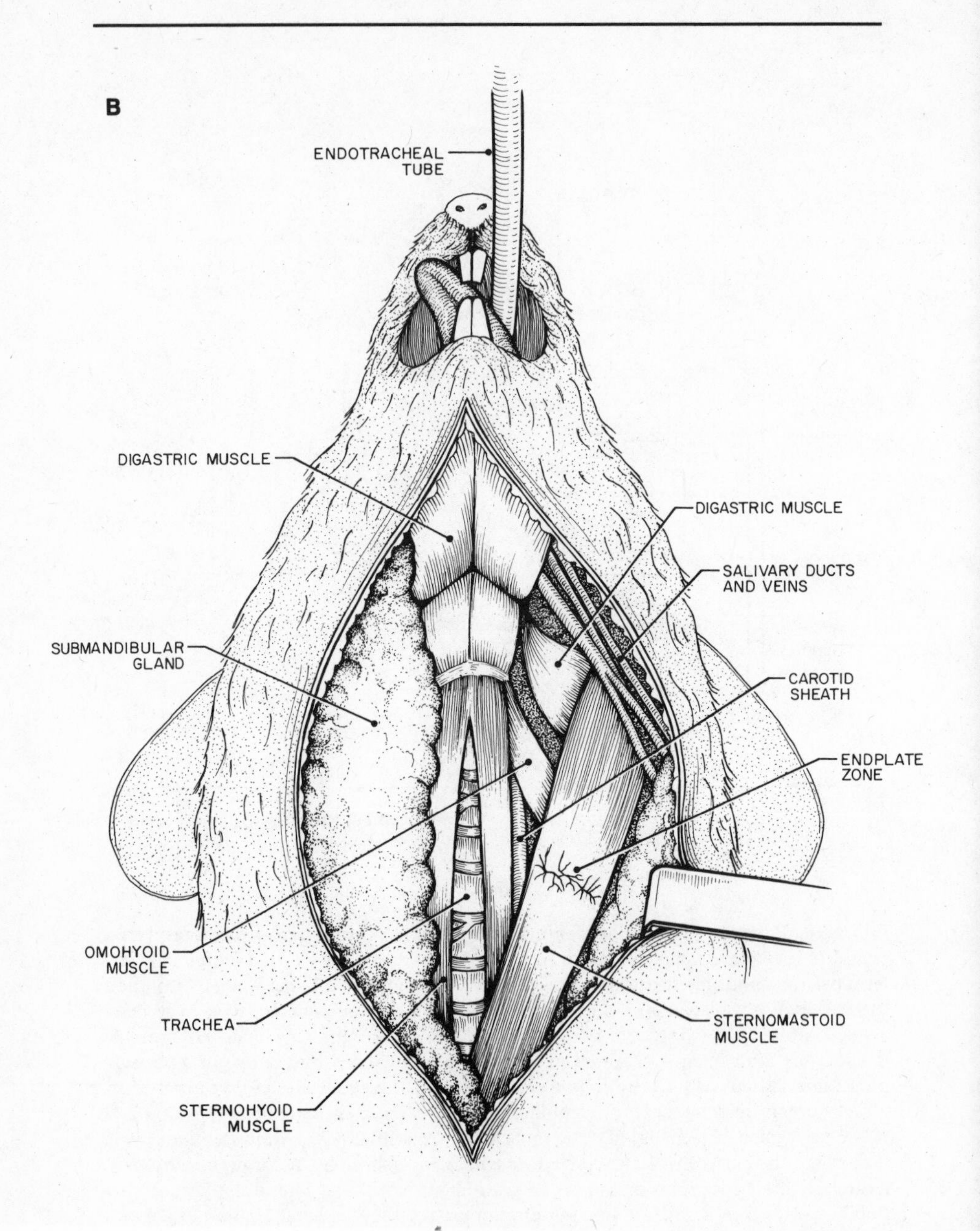
B
ENDOTRACHEAL TUBE
DIGASTRIC MUSCLE
DIGASTRIC MUSCLE
SALIVARY DUCTS AND VEINS
SUBMANDIBULAR GLAND
CAROTID SHEATH
ENDPLATE ZONE
OMOHYOID MUSCLE
TRACHEA
STERNOMASTOID MUSCLE
STERNOHYOID MUSCLE

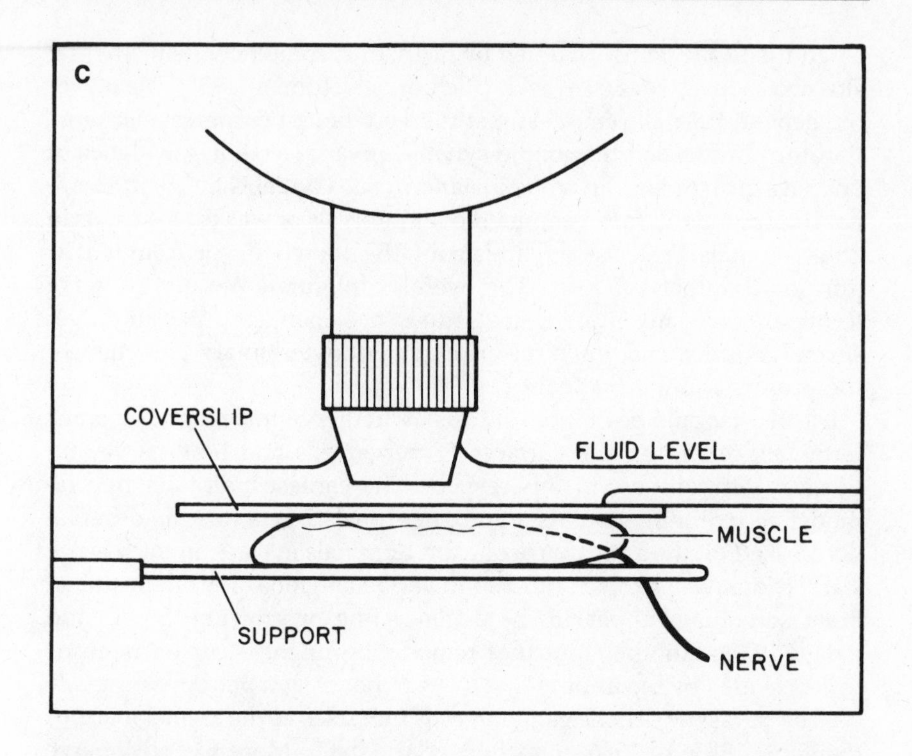
C
COVERSLIP
FLUID LEVEL
MUSCLE
SUPPORT
NERVE

D

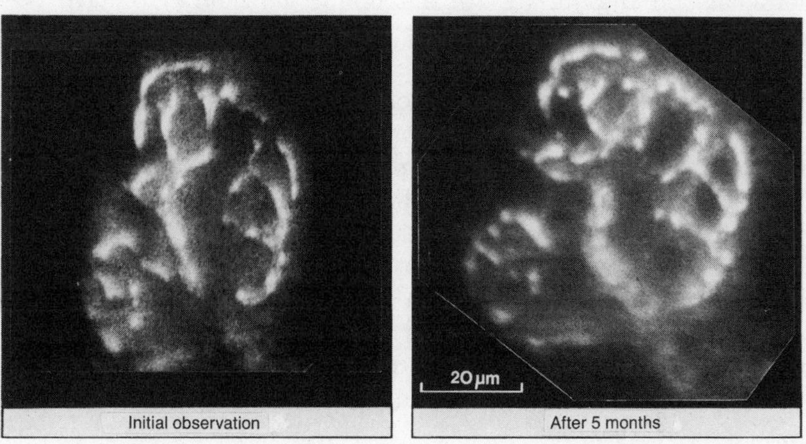
20 μm
Initial observation
After 5 months

an endplate site being occupied by *two* axons, some remodeling of that site occurs over several weeks (Rich and Lichtman, 1986). Whatever the general rules of remodeling at the vertebrate neuromuscular junction turn out to be, this simple synapse invariably shows evidence of ongoing growth and, in some instances, substantial changes in terminal configuration when examined over intervals of weeks or months in adult animals. Yet the dynamism of the vertebrate neuromuscular junction is rather subdued. The overall configuration of motor nerve terminals on many muscle fibers does not change appreciably over several months, and when the arrangement of terminals does change, the process is quite gradual.

Clearly it would be of interest to ask whether remodeling of synaptic terminals occurs on the surfaces of nerve cells and how nerve and muscle cells compare in this respect. The earliest indication that remodeling of endings occurs among neuronal targets was the observation, based on the appearance of axon terminals in electron microscopical sections, that axonal branches in various regions of the mammalian brain sometimes appear to be degenerating or growing (Sotelo and Palay, 1971). Another hint that remodeling might occur on neurons was provided by the finding that the number of synaptic boutons made on frog parasympathetic ganglion cells increases as the animal matures (Sargent, 1983a,b, 1986; Streichert et al., 1987). More recently, nerve terminals on the surfaces of autonomic ganglion cells have been studied prospectively by means of vital staining, low-light level videomicroscopy and digital image processing (Purves et al., 1987). In these experiments the problems of observing the much smaller and more complex terminals on neurons are formidable. The difficulties are minimized, however, by following the pattern of synaptic endings on the cell bodies of ganglion cells that lack dendrites, which provide a postsynaptic surface that is large and relatively easy to image. When such neurons are monitored over periods of several weeks in adult mice, the pattern of synapses on the surface of many cells changes appreciably (Figure 6.7). Perhaps the greater rate of remodeling on these ganglion cells compared to fibers in mouse muscle is related to the fact that many of these neurons are innervated by more than one axon (Snider, 1987).

The dendrites of sympathetic neurons in mature mice—that is, the postsynaptic rather than the presynaptic branches—have also been found to undergo substantial remodeling (Purves and Hadley, 1985; Purves, Hadley, and Voyvodic, 1986). In these studies the dendrites of identified nerve cells in the superior cervical ganglion of adult mice

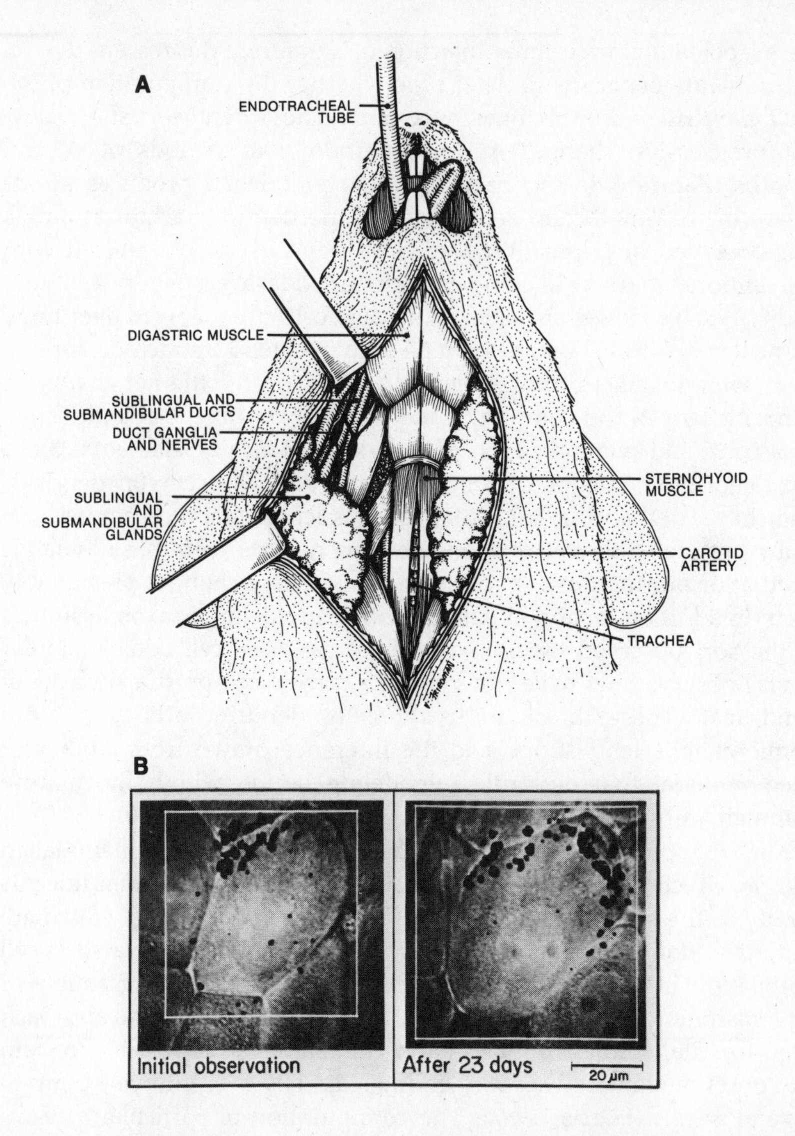

Figure 6.7. Remodeling of axon terminals on the somata of identified parasympathetic ganglion cells in an adult mouse. (*A*) The parasympathetic (salivary duct) ganglia. Synaptic terminals are stained by intravenous injection of a styryl pyridinium dye, and a map is made of the preganglionic endings on the surface of identified neurons. (*B*) The synaptic pattern initially and 23 days later, indicating that over this period the arrangement of axon terminals on this neuron has changed. The same neuron can be reidentified at the second observation by virtue of its size, shape, neighbor relationships, and position in the ganglion. Although on this particular neuron the number of endings deployed on the cell surface has increased over the interval, the change could as readily have been a loss of terminals (or simply a change in position without any change in number). (*A* from Purves and Lichtman, 1987; *B* from Purves et al., 1987)

were labeled by intracellular injection of a nontoxic fluorescent dye on two separate occasions to determine whether the configuration of the cell's dendritic arbor changes over time. These dendrites show slow but progressive changes in configuration over periods of several months (Figure 6.8). On balance, such remodeling produces an increase in the length and complexity of the dendritic arborization, as was observed in population studies of dendritic arbors labeled with conventional markers like horseradish peroxidase (see Figure 4.5; Voyvodic, 1987b). However, when individual cells are followed over time, particular dendritic branches can be seen to extend, retract, disappear, or in some instances form *de novo*. In other words, the net change in dendritic length and complexity that occurs over time is accomplished by a continual reorganization of individual dendritic elements. Since the majority of synapses in sympathetic ganglia occur on dendritic branches (Forehand, 1985), synapses on such ganglion cells must normally undergo a degree of change commensurate with these dynamic fluctuations of the postsynaptic dendrites. These changes presumably occur in addition to, and in parallel with, remodeling of axon terminals of the sort observed on the surfaces of ganglion cell bodies (similar *in vivo* observations have not yet been carried out on the surfaces of dendrites). These direct observations of dendritic arbors in autonomic ganglia lend support to the inference drawn from studies of fixed material that dendritic remodeling occurs within the mature mammalian brain (Weiss and Pysh, 1978; Berry et al., 1980).

A more rapid form of terminal remodeling occurs in the mammalian eye, in which axons from sensory neurons in the trigeminal nucleus ramify in the surface of the cornea. Corneal epithelial cells, which are subject to daily wear and tear, turn over in about 7 days; a stem cell population in the deeper layers of the cornea serves as a source of replenishment (Hanna and O'Brien, 1960). The same general approach used for vital studies in muscle and ganglia makes it possible to stain and observe identified sensory endings directly over intervals of up to several weeks (Harris, 1987). The configuration of particular sensory terminals does not change appreciably over a few hours. After a few days, however, the arrangement of endings indicates extensive remodeling. The reconfiguration of these sensory axon branches may be indicative of remodeling that occurs generally among terminals that ramify in surface epithelia. More important, these findings show that anatomical reorganization of neuronal branches can happen quite quickly.

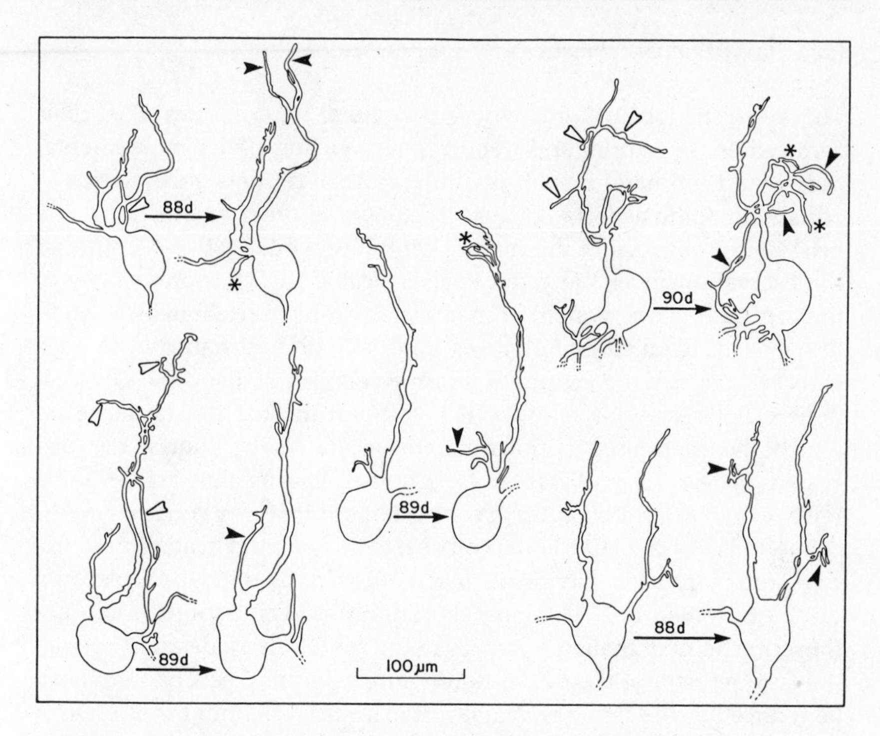

Figure 6.8. Ongoing changes in the dendritic geometry of individual sympathetic ganglion cells in living mice. Dendrites are reconstructed from images of an identified neuron filled with a nontoxic fluorescent dye at the beginning (left member of each pair) and the end (right member) of a 3-month period of study (the actual number of days that elapsed between the two images is indicated in each case). Black arrows indicate branches that have extended, white arrows indicate branches that have retracted, and asterisks indicate branches that appeared *de novo* in the interval. (From Purves, Hadley, and Voyvodic, 1986)

All told, these observations in adult mammals make three broad points: neuronal connections in maturity undergo obvious remodeling if provoked to do so by various experimental manipulations; these elicited responses are generally similar to the regulation of axonal and dendritic branches observed in development; and various degrees of physiological remodeling occur under normal circumstances in those parts of the nervous system in which it has been possible to follow the axonal and dendritic processes of identified neurons over substantial intervals.

Uncertainties about Plasticity of Central Connections

The apparent facility with which peripheral neurons can alter their connections by axonal and dendritic remodeling under experimental duress and, in many instances, under normal circumstances seems at odds with some aspects of what is known by other means about the behavior of neurons in the mature central nervous system. A number of observations over the years have indicated that the connectivity of the central nervous system of mammals or other vertebrates—even at the level of detailed circuitry—is quite *inflexible* in maturity, though early plasticity in the central nervous system is not disputed. Older, if indirect, observations relevant to this point include the evidence on which the traditional notion of hard-wiring in the central nervous system is based (see Chapter 1). Another finding that argues for a relatively static arrangement of connections in maturity concerns "critical periods" in the central nervous system. The modification of structure, function, or behavior during maturation is often possible only during a limited time. The most thoroughly studied critical periods are those in the mammalian visual system. Although modest changes in the arrangement of the ocular dominance columns (see Chapter 5) can be induced in the monkey visual cortex for up to 14 months after birth (LeVay et al., 1980)—and for even longer periods when using more subtle criteria, such as the transmitter content of neurons (Hendry and Jones, 1986) or various psychophysical tests (Harwerth et al., 1986)—the neural connections in this part of the brain appear to be largely fixed in adult animals. Thus, removal of an eye, or eyelid closure, does not produce the dramatic changes that these procedures elicit at birth (Hubel et al., 1977). Even in humans, the plasticity of the primary visual system appears to persist for only a few years. If a congenital strabismus is corrected by age 2 or 3, for example, the child grows to maturity with normal vision; however, such children are left with a permanent impairment, called amblyopia, if the misalignment of the eyes is corrected after this age (Von Noorden, 1980; Boothe et al., 1985). Similarly, studies of the development of the auditory system in birds (Knudsen, Esterly, and Knudsen, 1984; Knudsen, Knudsen, and Esterly, 1984; Knudsen, 1985a) and the somatosensory system in rodents (Belford and Killackey, 1980; Woolsey, 1984) indicate a progressive reduction in neural plasticity as these animals gradually attain maturity.

In spite of this unimpeachable evidence for a rapid "hardening" of connections in several developing sensory systems, other evidence suggests that neural connections in some parts of the vertebrate brain remain quite malleable in adulthood. One example is the mammalian primary somatosensory cortex (Kaas et al., 1983). When the arrangement of peripheral targets is perturbed in adult animals by amputation or nerve section, the cortical representation of the body undergoes substantial reorganization. These changes include expanded and even duplicate representation of the remaining parts (adjacent fingers, for example, after amputation of a digit). The mechanism of this phenomenon is not yet known. Some aspects of the reorganization occur so rapidly that a functional rather than anatomical basis is likely. Other aspects of such reorganization are quite slow, and are therefore compatible with altered patterns of connections.

Another case in point is the changing connectivity in the brains of some adult songbirds. Many birds have elaborate songs that are species-specific. Field studies have shown that, for most of these species, song is learned from conspecifics during a limited time in early life; indeed, one of the first uses of the term "critical period" was in this context (Thorpe, 1961; Marler, 1981; Nottebohm, 1984). However, some species, such as the canary, continue to embellish their repertoire from season to season when, by other criteria, they are fully mature (Nottebohm and Nottebohm, 1978). Related studies have established the location, size, and connectivity of the song control centers as well as the morphology of the cells in these loci (Nottebohm, 1980). The song-control nuclei of birds with more extensive repertoires are larger than those of less talented peers and of the female of the species, which does not usually sing (Nottebohm et al., 1981; Brenowitz and Arnold, 1986). Moreover, the overall size of these nuclei fluctuates from season to season. Thus, the brain nuclei related to song are larger in the breeding season and smaller during the period when the birds are sexually inactive (Gurney and Konishi, 1980; Devoogd and Nottebohm, 1981a,b; Gurney, 1981; Nottebohm, 1981). These fluctuations are assumed to be the result of changes in neuronal number and in the extent of neuronal arborizations, which are generated in turn by different levels of sex hormones (Arnold, 1985). This interpretation has been strengthened by the finding that the dendritic length and complexity of the spinal motor neurons in the rat which innervate the perineal musculature are influenced by androgen levels (Forger and Breedlove, 1986; Kurz et al., 1986). In some seasonally breeding rodents, the size and form of the perineal muscles actually change according to the time of year. The implication of this seasonal variation in sexual musculature is that hormonally induced changes in the targets generate changes in the form of the relevant motor neurons and, therefore, of synaptic connectivity (Forger and Breedlove, 1986).

A reasonable interpretation of these various studies on the malleability of the mature central nervous system would be that the brain retains a considerable capacity for connectional change, although this capacity evidently diminishes with age. Indeed, in some regions of the brain plasticity of connections cannot be demonstrated beyond early life. This diminished ability of neural connections to change is generally in keeping with the clinical observations that recovery from neural injury is more rapid and complete in young children (or young experimental animals) than in adults (Kennard, 1936, 1942) and that the capacity to learn decreases with age. These ideas also accord with events in the peripheral nervous system, where the vigor of trophic interactions and the various rearrangements they engender diminish with age. In spite of the diminished overall capacity for connectional change that is evident with age, many and perhaps all parts of the nervous system retain some ability to alter their synaptic connections. Moreover, in those instances in which direct study has been possible in physiological circumstances, ongoing changes in the configuration of synaptic connections have been observed.

Reasons for Ongoing Trophic Interactions

There is, of course, no clear boundary between development and maturity. These are simply convenient words used to distinguish, in a broad sense, early life from later life. Accordingly, there is no *a priori* reason to suppose that the developmental mechanisms for regulating neural connections which are apparent in early life cease to operate at some point.

There are several reasons why the continuing operation of trophic mechanisms might be useful. First, most mammals and other vertebrates mature over a long period during which the size and form of the body continue to change. In many mammals (rats and guinea pigs provide well-studied instances), growth persists well beyond the time of sexual maturity. As a result, the innervation of changing targets and the configuration of neural maps must be capable of coordinated modification in an ongoing manner. Still later in life, the body undergoes further changes as various tissues express the physiological atrophy associated with aging. Thus, the nervous system must adapt to the further somatic changes imposed by senescence.

A second rationale for the persistence of trophic mechanisms in maturity is the need to compensate for the continual wear and tear to which bodies are subjected. This phenomenon is most apparent at body surfaces, where various classes of sensory receptors are subject to physiological damage, and in tissues like muscles, which suffer significant trauma during normal use. All but the most sedentary readers will be familiar with the considerable pain that occurs the day after strenuous exercise. This syndrome is actually the result of tissue damage. Thus, in man, postexercise studies of conditioned athletes show elevated levels of serum enzymes indicative of tissue destruction, myogloburia (a specific indicator of muscle damage), and a variety of structural changes in muscle biopsies, including disruption of myofibrils, muscle fiber necrosis, and local inflammation (Fridén, 1983, 1984; Fridén et al., 1983; Hikida et al., 1983). Similar changes have been observed in mouse muscles after exercise (Salminen and Vihko, 1983; McCully and Faulkner, 1985; Irintchev and Wernig, 1987). The ability to repair such physiological damage is probably a feature of many tissues, including neural tissues. Similarly, the nervous system must cope with the effects of disease and other more obvious sorts of injury. Poliomyelitis provides an instructive example. Electromyographic studies show that the spinal motor neurons that survive an initial infection with polio virus form much larger motor units in the muscles they innervate (Cashman et al., 1987). This effect is presumably explained by sprouting of the residual neurons to innervate muscle fibers left denervated by the death of some motor nerve cells during the acute stages of the disease. In old age, some victims of childhood polio develop a syndrome of weakness that apparently reflects the inability of overextended motor neurons to maintain innervation of their full complement of target cells—a finding in keeping with the general idea that the vigor of trophic interactions wanes with age and that adjustments to atrophic changes are necessary in normal aging.

A third, quite speculative, reason why ongoing trophic interactions may be useful in maturity is the persistent need to encode information in the nervous system. The heritage of psychology in this century has tended to make us think of learning as a phenomenon concerned with relatively short-term associations (as in classical and operant conditioning), problem solving, and the capacity to retain information briefly presented. However, for the sorts of learning that transpire almost imperceptibly over much longer periods and that are not readily lost—such as the gradual learning of a novel and complex motor skill, the acquisition of a new language, or even the gradual adaptation to the bifocal lenses that for many of us mark the undeniable onset of middle age—the ongoing malleability of neural connections may be crucial. For all these reasons, the persistence of neural adjustment by means of neuron-target interactions is a useful, and probably essential, property of the mature, as well as the developing, nervous system.

SEVEN

A Molecular Basis for Trophic Interactions in Vertebrates

THE IDEA that the changing size and form of vertebrates requires feedback between neurons and the cells they innervate—and that these adjustments involve, at the cellular level, ongoing regulation of axonal and dendritic branches—does not stand or fall on the knowledge of any particular signal that mediates these coordinated interactions. Nevertheless, at least one well-studied molecule, nerve growth factor (NGF), appears to be precisely the sort of intercellular signal required to generate many of the adjustments of neural organization considered in the preceding chapters. The further fact that one of the major classes of neurons sensitive to the action of NGF is the same group of autonomic (sympathetic) ganglion cells that figures prominently in the foregoing discussion of axonal and dendritic regulation makes this molecule all the more pertinent. Extending the context for trophic agents like NGF from neuronal survival to the regulation of neuronal form and connections provides a realistic molecular mechanism for many aspects of the trophic control of innervation evident in ontogeny and phylogeny.

NGF as a Regulator of Neuron Survival

Nerve growth factor was discovered in the late 1940s and early 1950s in the course of research aimed at understanding the mechanism of neuronal survival during embryonic development. A disagreement had arisen about interpretation of the reduced number of sensory and motor neurons following limb ablation in the chick embryo. V. Hamburger and others had argued that the depletion of nerve cells after limb ablation represented a failure of differentiation, whereas R. Levi-Montalcini and G. Levi correctly surmised that the cause was neuronal degeneration (Hamburger, 1934; Levi-Montalcini and Levi, 1942, 1944). In order to resolve the matter, Levi-Montalcini came to Hamburger's laboratory in St. Louis in 1947 for what was to have been a one-year stay. During the initial period of their collaboration, Hamburger and Levi-Montalcini determined that limb bud ablation did indeed cause degeneration of neurons in spinal ganglia and, more important, that the death of some neurons was a normal developmental event (Hamburger and Levi-Montalcini, 1949). They further suggested that targets must provide a signal of some sort to the relevant neurons and that a limited amount of such a signal might explain the competitive nature of cell death (see Chapter 3). In the light of these hypotheses, Levi-Montalcini (who had abandoned the idea of an early return to Italy) and Hamburger undertook a series of experiments to explore the source and nature of the postulated signal (Levi-Montalcini and Hamburger, 1951). In the course of this work, they became aware of a remarkable result that had been obtained by E. Bueker (1948), a former student of Hamburger's. Bueker had removed a limb from chick embryos and implanted a piece of mouse tumor in its stead; his purpose was to determine whether a histologically homogeneous neoplasm could substitute for the limb in maintaining the related nerve cells in the spinal cord and sensory ganglia (Bueker, 1985). The outcome was that the tumor appeared to furnish the same stimulus as a limb and, furthermore, caused an appreciable enlargement of the sensory ganglia that normally innervate the appendage (Bueker, 1948). Intrigued by this result, Levi-Montalcini and Hamburger asked Bueker if they might pursue the basis of the tumor's effect. He consented, and in a series of experiments carried out between 1949 and 1953, Levi-Montalcini and Hamburger provided evidence that Bueker's tumor (mouse sarcoma 180) secreted a soluble factor that stimulated the survival and growth of sensory and sympathetic ganglion cells (Levi-Montalcini and Hamburger, 1951, 1953; Levi-Montalcini, 1953).

Levi-Montalcini then devised a bioassay for the presumed agent, which had been named nerve growth factor, and, in collaboration with S. Cohen, set about isolating and characterizing the NGF molecule (Cohen et al., 1954; Levi-Montalcini et al., 1954). NGF was identified as a protein and substantially purified from a rich biological source, the salivary glands of the male mouse (Cohen and Levi-Montalcini, 1956; Levi-Montalcini and Cohen, 1956; Cohen, 1959, 1960). Parenthetically, the reason for so much NGF in these glands is a mystery; its purpose there has little or nothing to do with innervation since the NGF is sequestered in the acini of the glands and secreted in the saliva. By 1971 NGF had been purified to homogeneity, and its amino acid sequence determined (Angeletti and Bradshaw, 1971; Angeletti et al.,

1973); by 1987 the genes encoding messenger RNA for NGF had been cloned in several species (see Levi-Montalcini, 1987). The biologically active component of the molecule, called β-NGF, is part of a larger complex, called 7S-NGF, the chemistry of which is now fairly well understood (Greene and Shooter, 1980).

The dramatic influence of NGF on cell survival (Figure 7.1A), together with what was known about the significance of neuronal death in development, suggested one biological function for this agent: to act as a target-derived signal that in the course of events helps to match the number of nerve cells to the number of target cells. Support for the idea that NGF is important for neuron survival in more physiological circumstances emerged from a number of further observations. Perhaps most important, when developing rodents are deprived of NGF by the chronic administration of an NGF antiserum, they grow to maturity lacking the great majority of sympathetic neurons (Figure 7.1B; Cohen and Levi-Montalcini, 1956; Levi-Montalcini and Cohen, 1956; Cohen, 1960; Levi-Montalcini and Booker, 1960; Levi-Montalcini, 1972). As a result of NGF deprivation, the normal death of a portion of sympathetic ganglion cells proceeds to the extinction of very nearly the entire population. Second, injection of exogenous NGF into newborn rodents causes a marked hypertrophy of sympathetic ganglia, an effect opposite that of NGF deprivation (Levi-Montalcini and Booker, 1960). When sympathetic ganglia are examined histologically or in the electron microscope after a period of NGF treatment, the neurons are both more numerous and larger; there is also more neuropil between cell bodies, suggesting an overgrowth of axons, dendrites, and other cellular elements (Hendry and Campbell, 1976; Schäfer et al., 1983).

Later work confirmed that NGF has little or no effect on many other classes of nerve cells and that its action is therefore quite specific. The effects of NGF, however, are not limited to sensory and sympathetic neurons in the peripheral nervous system. NGF and NGF receptors have been found in specific regions of the rodent brain, and NGF is now known to affect some classes of central neurons in much the same manner as peripheral ganglion cells (Seiler and Schwab, 1984; Gage et al., 1986; Korsching et al., 1985; Korsching, 1986; Large et al., 1986; Shelton and Reichardt, 1986; Williams et al., 1986; Auburger et al., 1987; Barde et al., 1987). For some years it remained puzzling why dorsal root ganglion cells, which are clearly sensitive to NGF by bioassay, are much less affected than are sympathetic neurons by a postnatal course of NGF antiserum. In fact, no obvious influence of NGF deprivation on sensory ganglion cells was observed at first. This mys-

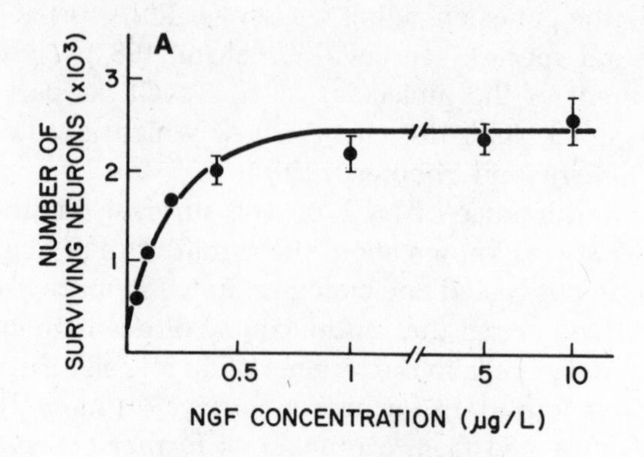

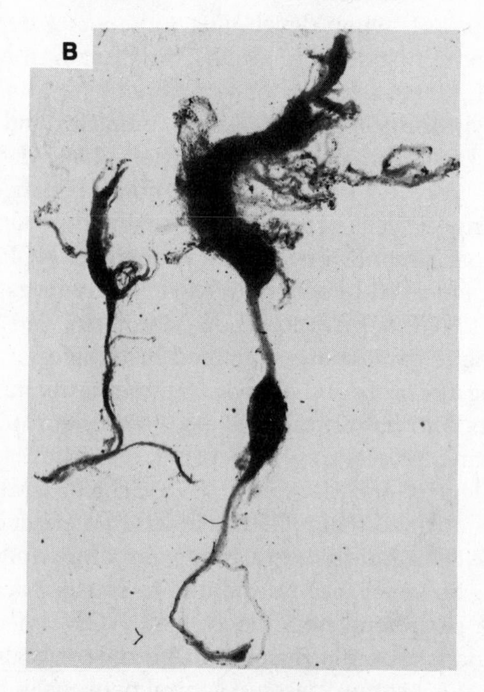

Figure 7.1. Effect of NGF on the survival of sympathetic ganglion cells. (*A*) The survival of newborn rat sympathetic ganglion cells grown in culture for 30 days evaluated quantitatively as a function of NGF concentration. Dose-response curves such as this one confirm the strict dependence of these neurons on the availability of NGF. (*B*) The stellate and sympathetic chain ganglia from a normal 20-day-old mouse (right) and the same ganglia from a littermate treated daily from birth with injections of NGF antiserum (left). The ganglia from the treated animal are markedly atrophic because, in the absence of NGF, the majority of the ganglion cells die. (*A* after Chun and Patterson, 1977; *B* from Levi-Montalcini, 1972)

tery was resolved by the discovery that, if one provides antiserum throughout early embryonic life by immunizing animals which are then made pregnant, a severe depletion of dorsal root ganglion cells develops (Gorin and Johnson, 1979; Johnson et al., 1980). Evidently, the NGF-sensitive period for dorsal root ganglion cells occurs earlier and is briefer than the sensitive period for sympathetic ganglion cells.

The ability of NGF to support neuronal survival (and of NGF antiserum to enhance cell death) is not in itself unassailable proof of a physiological role for this protein in development. For instance, these observations provide no direct evidence of NGF synthesis by (and uptake from) neuronal targets. This gap was filled by a further series of ingenious experiments at the cellular and molecular levels. Studies using radio-labeled NGF showed that the exogenous protein, when injected into the eye, is taken up by specific receptors at nerve endings and is returned to the cell bodies of sensitive neurons in the superior cervical ganglion by retrograde axonal transport (Thoenen and Barde, 1980). It is still unclear, however, whether the relevant signal acting within NGF-sensitive cells is the NGF molecule itself, which is known to be internalized together with NGF receptors, or a second messenger (Palmitier et al., 1984). NGF was also shown to be present in sympathetic targets and to be quantitatively correlated with the density of sympathetic innervation (Korsching and Thoenen, 1983). Finally, messenger RNA for NGF was demonstrated in targets innervated by sympathetic and sensory ganglia, but not in the ganglia themselves or in targets innervated by other types of nerve cells (Heumann et al., 1984; Shelton and Reichardt, 1984; Korsching and Thoenen, 1985; Davies et al., 1987). In other words, the targets of NGF-sensitive neurons synthesize NGF. Functional tests in intact animals corroborated the physiological significance of target synthesis. Thus, cutting the axons of NGF-sensitive neurons, such as sympathetic ganglion cells, causes them to die. Conversely, neuronal death after axotomy or normally occurring neuronal death can be prevented by supplying exogenous NGF (Hendry, 1975; Hendry and Campbell, 1976; Hamburger and Yip, 1984).

In sum, several decades of work in a number of laboratories have shown that NGF meets all the criteria for a trophic factor that mediates cell survival among two specific neuronal populations in birds and mammals (sympathetic and neural crest-derived sensory ganglion cells). These criteria include: the death of the relevant neurons in the absence of the factor; the survival of a surfeit of neurons in the presence of the factor; the presence of the agent in the neuronal targets; the synthesis of the factor by target cells; the presence of specific receptors for the agent on the innervating nerve terminals; and the retrograde axonal transport of the agent (or a second messenger) to convey the target-derived signal to the neuronal cell body.

NGF as a Regulator of Neuronal Processes

In addition to mediating the survival of specific classes of nerve cells, NGF influences the neuritic processes of the surviving nerve cells. It is this property of NGF that is particularly important to the general argument that much neural adjustment to somatic change occurs by the rearrangement of axonal and dendritic branches and their synaptic connections.

Prima-facie evidence for this further action of NGF comes from the fact that explanted sensory or sympathetic ganglia exposed to a culture medium containing NGF show a marked outgrowth of neurites within 24 hours (Figure 7.2). Indeed, this response is the basis for the bioassay first described by Levi-Montalcini and her collaborators in the early 1950s (Levi-Montalcini et al., 1954). In spite of the widespread use of the neurite-outgrowth bioassay, relatively little attention was paid to this aspect of NGF's influence during the period when workers were attempting to clarify its role as an agent of neuronal survival. One difficulty in assessing the influence of NGF on neuronal branching is the need to dissociate its specific effects on neurites from effects that are secondary to the "good health" that sensitive neurons enjoy in its presence. This problem was solved by the use of a novel culture system that distinguished the local effects of NGF on neurites from effects mediated through the cell body (Campenot, 1977, 1981, 1982a,b). In this system, dissociated sympathetic ganglion cells from young rats are placed in the central well of a chamber with three compartments (Figure 7.3). Each compartment is isolated from the others by seals on the bottom of the culture dish, a design which allows the NGF concentration to be varied independently in the several chambers. If all three compartments contain medium with adequate concentrations of NGF, then neurites from the ganglion cells in the central well extend through the seals into the peripheral compartments. However, if NGF is removed from one of the peripheral wells, then the neurites that have grown into that compartment gradually retract. Conversely, if NGF is removed from the central compartment but retained in the peripheral wells, the neurites remain in place (Campenot, 1982b). These results, in agreement with earlier work on NGF as an agent of neuronal sur-

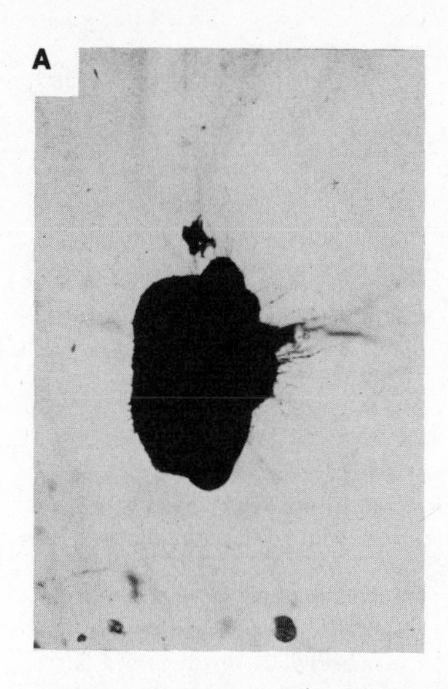

Figure 7.2. Effect of NGF on the outgrowth of neurites. *(A)* A chick sensory ganglion taken from an 8-day embryo and grown in organ culture for 24 hours in the absence of NGF. Few, if any, neuronal branches grow out into the plasma clot in which the explant is embedded. (*B*) A similar ganglion in identical culture conditions 24 hours after the addition of NGF to the medium. NGF stimulates a halo of neurite outgrowth from the ganglion cells. Similar results are obtained with sympathetic ganglia. (Courtesy of R. Levi-Montalcini)

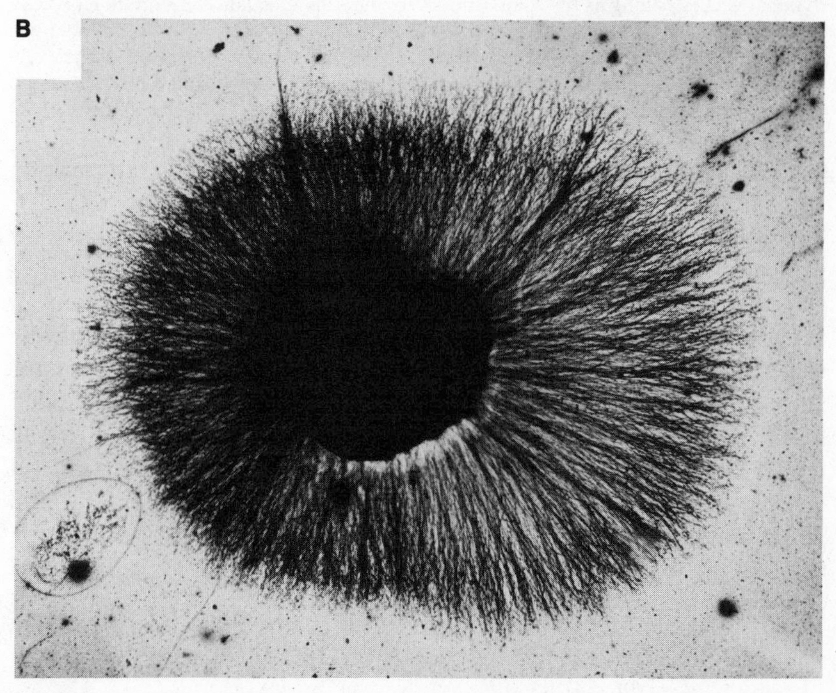

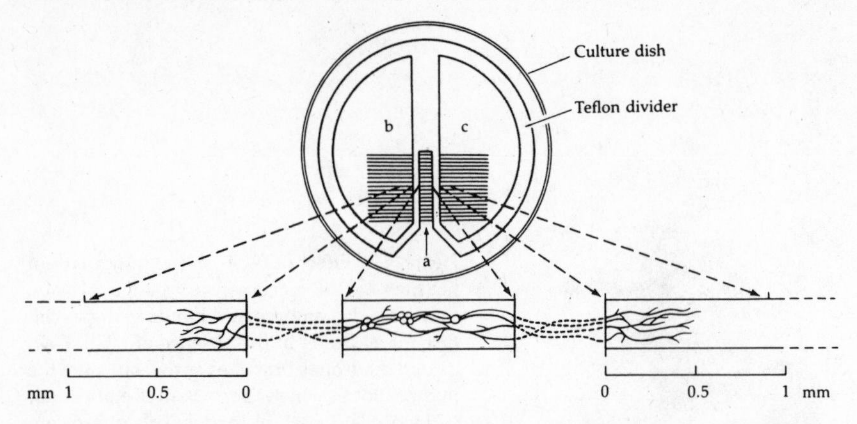

Figure 7.3. Evidence that NGF can influence neurite growth by local action. Three sections (*a,b,c*) of a culture dish are separated from one another by a Teflon divider sealed to the bottom of the dish with grease (the view is looking down on the dish). Isolated rat sympathetic ganglion cells plated in compartment (*a*) can grow along avenues of collagen on the surface of the dish, through the grease seal, and into compartments (*b*) and (*c*). (An expanded view in the same orientation is shown below.) Growth into a lateral chamber occurs as long as the compartment contains an adequate concentration of NGF. Subsequent removal of NGF from a compartment causes a local regression of neurites, without affecting the survival of cells or neurites in the other compartments (when NGF is absent in the peripheral compartments, it must be present in the central chamber, since the neurons need it to survive). Neuritic growth can thus be locally controlled by a trophic agent. (After Campenot, 1981)

vival, show that NGF can maintain nerve cells by retrograde transport from the "periphery." More importantly, they indicate that neurites extend or retract as a function of the *local* concentration of this agent. If the effects of NGF on neurites were mediated primarily through the cell body, then the neuronal branches in different compartments would respond similarly. In short, neuronal branches can be controlled locally by trophic stimuli and do not simply depend on the overall effects of trophic agents on the parent cell (see also Gunderson and Barrett, 1980). As a result, some branches of a neuron may extend while others retract, which is of course what actually happens during the various experimental and physiological rearrangements of neural connections described in Chapters 5–6.

Effects of NGF on Axons, Dendrites, and Synapses

Because axons and dendrites are not easily distinguished in culture, *in vitro* experiments do not yield information about the trophic effects of

NGF on these two types of neuronal branches. Work carried out on intact animals, however, has shown that NGF affects both the axons and the dendrites of sympathetic ganglion cells, albeit in different ways. This work has also shown that NGF influences the maintenance of synapses on ganglion cells.

The introduction in the 1960s of a specific stain for sympathetic (catecholaminergic) axons, the Falck-Hillarp technique, furnished a means of directly studying the regulation of axonal growth in sympathetic targets such as the iris (Olson and Malmfors, 1970). With this method it was possible to show that treatment of rodents with exogenous NGF stimulates a proliferation of sympathetic axons in the target tissue (Olson, 1967). Similarly, if the iris is transplanted from a donor rat into the anterior chamber of the eye in a host animal, over a period of 1–2 weeks the auxiliary iris becomes innervated by about the normal density of sympathetic axon terminals (Figure 7.4; Olson and Malmfors, 1970). The innervation of the implant is maintained indefinitely. A supernumerary target thus elicits axonal sprouting in maturity and continues to support the innervation it receives. This result implies a diffusible signal from the transplanted iris which stimulates renewed

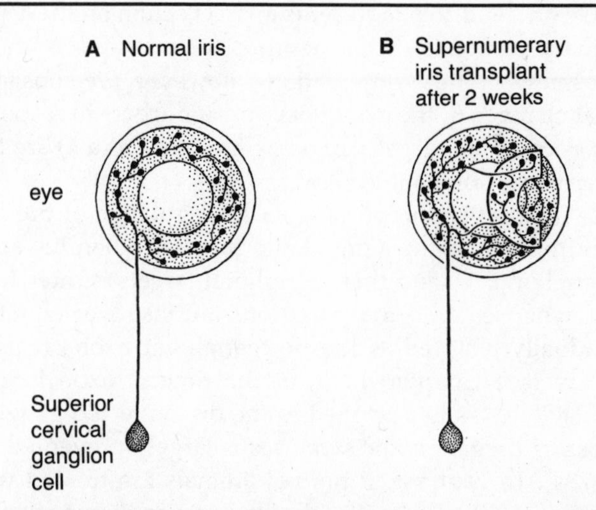

Figure 7.4. Response of sympathetic ganglion cell axons to ocular implantation of a supernumerary iris in the rat. (*A*) Normal innervation of the iris by a superior cervical ganglion cell. (*B*) Within 2 weeks of implantation of an iris from a donor animal into the anterior chamber, the host axons sprout new branches which innervate the additional tissue. The density of innervation in the host iris also increases in response to the stimulus presented by the implant. There is strong circumstantial evidence that the molecular signal for this effect is NGF. (After Olson and Malmfors, 1970)

growth of the sympathetic axons that already innervate the host iris. The difference between this result and other experiments on sprouting—in skeletal muscle, for instance—is that NGF can be implicated (see also Diamond et al., 1987). Because the isolated iris is known to synthesize NGF (Heumann and Thoenen, 1986; Shelton and Reichardt, 1986) and because NGF is known to stimulate neuritic outgrowth from sympathetic ganglia, it is reasonable to conclude that this molecule constitutes at least part of the stimulus provided to sympathetic axons by the transplant. These observations, coupled with what is now understood about the biology of NGF, suggest a more specific interpretation of the sprouting response to partial denervation previously described (see Chapter 6): when targets are denervated, the increased availability of a target-derived trophic signal may stimulate nearby axons to renewed growth.

NGF has also been shown to have a marked effect on dendritic growth. Thus, when sympathetic ganglion cells from rat pups given daily injections of exogenous NGF are subsequently assessed by intracellular marking, the dendrites of neurons in the experimental animals are both longer and more numerous after two weeks of treatment (Figure 7.5; Snider, 1988; see also Schäfer et al., 1983). Whether the effect of NGF on dendrites represents a direct action on these processes or a secondary response of the neurons to retrogradely transported NGF is not known. These observations, however, are consistent with the idea that changes in the extent of dendritic arbors in response to an increase or decrease in target size (see Chapters 4 and 6) are the result of target-derived trophic molecules.

Finally, NGF has been implicated in the regulation of preganglionic innervation in mammalian sympathetic ganglia. When the axonal link between ganglion cells and their peripheral targets is interrupted, the majority of synapses onto these neurons are lost over about a week, but are gradually restored as the postganglionic axons regenerate to the periphery (see Chapter 6). If, at the time of axon interruption, exogenous NGF is locally supplied to the disconnected ganglion cells, then the loss of preganglionic synapses is largely prevented (Njå and Purves, 1978a). In contrast, if normal animals are treated with daily injections of an NGF antiserum, the loss of ganglionic synapses and other effects seen after postganglionic axotomy are elicited. These findings suggest that the loss of synapses after axotomy is also related to the availability of a trophic signal. In support of this interpretation, treatment of neonatal rats with exogenous NGF increases the number of ganglionic synapses (Schäfer et al., 1983).

The mechanism of NGF's effects on the maintenance of ganglionic synapses is more complex than, for example, the action of this trophic agent during sprouting and synaptogenesis in the iris. In that instance NGF acts directly on the postganglionic axons. The most likely explanation for synapse loss after axon interruption is that the diminished availability of target-derived NGF to ganglion cells affects in turn the availability of a *second* trophic signal that operates between ganglion cells and the preganglionic innervation they receive. Presupposition of

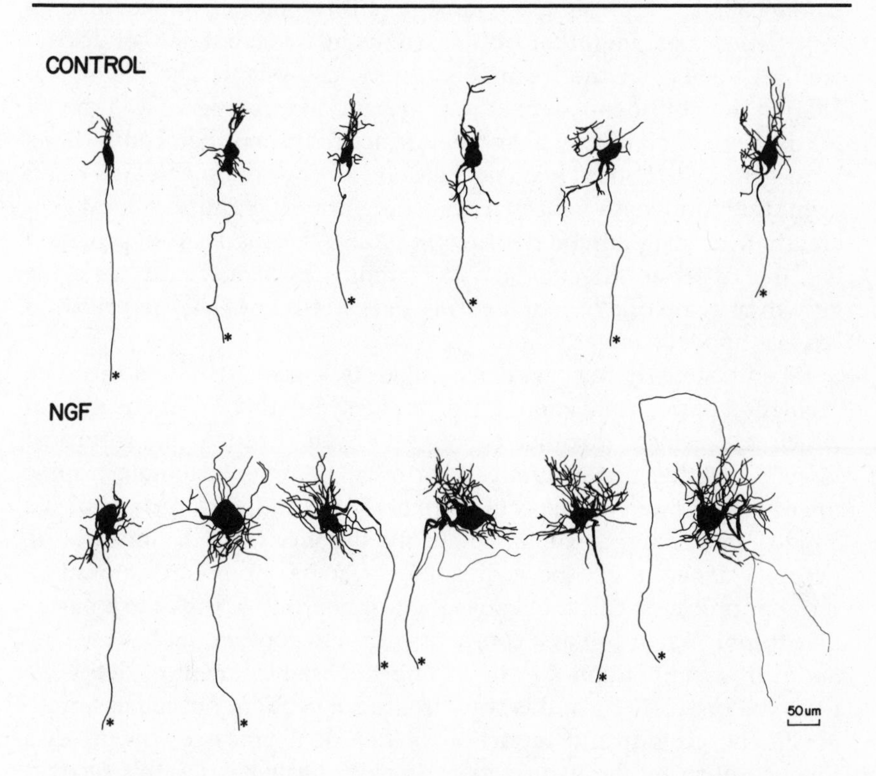

Figure 7.5. Effect of NGF on the dendritic development of superior cervical ganglion cells. Neonatal rats were given a 2-week course of daily, subcutaneous injections of NGF; individual neurons from treated and control animals were subsequently filled with the enzyme horseradish peroxidase by intracellular injection to assess their geometry. Neurons from the untreated control animals have shorter and less complex dendrites than ganglion cells from NGF-treated littermates. The number of primary dendritic branches extending from the cell body is also increased. Thus, the size of the dendritic arbors attained by neonatal animals after 2 weeks of NGF treatment is about the same as that attained by untreated animals after 3 months. Asterisks indicate the axon of each cell. (After Snider, 1988)

this tandem effect is necessary because ganglion cells do not themselves synthesize NGF, nor is there any evidence that preganglionic axons are sensitive to it.

A General Scheme for the Action of Trophic Molecules

NGF is generally accepted as the molecular agent of neuronal survival for two particular classes of sensory and sympathetic nerve cells. Because competition for survival is widely observed in neural pathways where NGF is *not* present, the biology of this molecule has come to be regarded as a paradigm for trophic agents inferred, but not yet discovered, elsewhere in the mammalian nervous system. The biology of NGF in the peripheral sympathetic system may also serve as a model for the regulation of axonal and dendritic arbors and their connections by target interactions. The major reasons for considering NGF to be an agent that influences neuronal branching are: NGF causes a local proliferation of sympathetic neurites *in vitro*, whereas NGF deprivation has the opposite effect; and NGF stimulates axonal and dendritic growth of sympathetic ganglion cells *in vivo*, whereas NGF deprivation has the opposite effect.

Taken together, this evidence suggests a general scheme for the regulation of neuronal connections in the sympathetic nervous system which may apply to other systems as well (Figure 7.6). Neuronal targets, whether non-neural cells or other neurons, elaborate trophic molecules in limited amounts. In embryonic and early postnatal life and to a more limited extent in maturity, the survival of the innervating neurons depends on the acquisition of some minimum amount of these trophic molecules. In consequence, neurons sensitive to a particular trophic agent initially compete with one another, and those that fail in this competition die. Following the establishment of definitive neuronal populations in this way, trophic dependency becomes apparent in the growth and retraction of neuronal processes, again as a function of target-derived support. In early postnatal life, this secondary dependence is evident in the rearrangement of initial connections whereby each neuron comes to innervate an appropriate number of target cells and each target cell comes to be contacted by an appropriate number of axons. In this way a suitable degree of convergence and divergence in the system is established. In general, a quantitatively appropriate distribution of innervation occurs because target cells that have received sufficient innervation present a diminished trophic stimulus to nearby axons, and axons that have already made sufficient contacts and are thereby receiving an adequate amount of trophic sup-

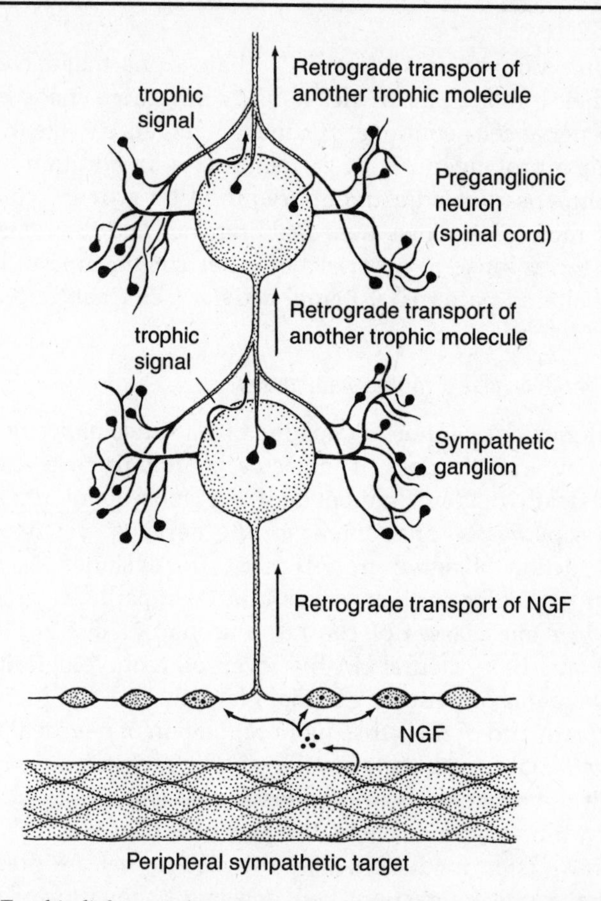

Figure 7.6. Trophic linkages taken to coordinate the innervation of targets with the innervation of central neurons. In the sympathetic nervous system of mammals, a trophic molecule, NGF, is elaborated by peripheral targets and interacts with specific receptors on the axon terminals of sympathetic ganglion cells. At this level, NGF modulates the branching of axon terminals by local action; increased availability of NGF causes the growth of terminals and decreased availability causes retraction. NGF (or a second messenger controlled by the interaction of NGF with its receptor) also affects ganglion cells in a more general way after uptake by axon terminals and retrograde transport to the parent cell body. The amount of trophic signal that reaches the cell body determines whether the neuron will survive, informs the neuron about the degree of its peripheral connectivity, and modulates the dendritic arbor of the ganglion cell. Because ganglion cells are in turn the target for preganglionic innervation that arises from neurons in the spinal cord, the elaboration of a second trophic molecule by ganglionic neurons is inferred (ganglion cells do not synthesize or secrete NGF). This second trophic signal controls the neurons that innervate ganglion cells in much the same way that NGF acts on the ganglion cells, namely by local action on the preganglionic terminals that innervate the ganglion cells and by more general retrograde effects on survival and dendritic growth after transport to the preganglionic cell bodies in the spinal cord. A further trophic linkage is postulated between the preganglionic neurons and the innervation they receive from higher autonomic centers. In this manner, information that initially derives from peripheral targets can affect the connectivity of an entire neural pathway. (After Purves, 1986)

port tend to compete less vigorously than axons that have not yet obtained their quota. Later in life, neural connections made by a fixed number of nerve cells continue to adjust as targets change in size and form during a prolonged period of maturation. In addition to mediating the compensatory adjustments required by growth, competition for trophic molecules allows neuronal branches and their connections to change in response to a variety of other circumstances, including injury and altered patterns of neural activity (see Chapter 8).

Other Target-Derived Trophic Molecules

Trophic interactions cannot be explained by a single molecule or, in all likelihood, by a single class of molecule. Although target-dependent nerve cell death in development and the subsequent regulation of neurites are pervasive phenomena in the nervous systems of vertebrates, the action of nerve growth factor, for instance, is limited to neurons in particular sorts of sensory and sympathetic ganglia (and, arguably, to some classes of central neurons). NGF has little or no effect on many other neural circuits, even on embryologically related systems like parasympathetic ganglia. Nevertheless, the prevalence of neuronal death and of the subsequent regulation of neuronal morphology suggests that functionally similar trophic molecules are important in many, if not all, regions of the nervous system.

Based on this logic, or at least the part of it related to cell survival, attempts have been made to isolate the factors that are presumed to mediate the trophic interactions between smooth muscle and parasympathetic ganglia (Ebendal et al., 1979; Bonyhady et al., 1980; Manthorpe et al., 1980; Nishi and Berg, 1981; Barde et al., 1983; Berg, 1984; Thoenen and Edgar, 1985; Thoenen et al., 1987; Watters and Hendry, 1985; Wallace and Johnson, 1987), between skeletal muscle and the spinal motoneurons that innervate them (Henderson et al., 1981; Gurney, 1984; Nurcombe et al., 1984; Bonyhady et al., 1985; Smith et al., 1985, 1986; Davies, 1986; Dohrmann et al., 1986; Hulst and Bennett, 1986; Oppenheim et al., 1987), and between nerve cells themselves in a variety of neural regions (Dribin and Barrett, 1980, 1982; Barde et al., 1982; Berg, 1984; Collins and Crutcher, 1985). Although no molecule has yet joined the ranks of NGF as an unambiguous example of a trophic factor, using criteria applicable to either survival or the regulation of neuronal processes, a spectrum of agents that may be analogous in function to NGF has been described.

The most advanced of these investigations is the purification of a factor from pig brain, called brain-derived neurotrophic factor, or BDNF (Barde et al., 1982, 1983; Barde, Davies, et al., 1987). This agent, which is a small basic protein, supports the survival and neuritic growth of primary sensory neurons from a variety of sources in the chick, including: neurons in the dorsal root ganglia, which are also sensitive to NGF; placodally-derived sensory ganglion cells, which are insensitive to NGF (Lindsay et al., 1985; Davies et al., 1986a); and retinal ganglion cells (J. E. Johnson et al., 1986). Understanding the biological role of BDNF has been difficult because of the large number of brains required to produce a significant amount of this agent and the low antigenicity of the molecule. Nevertheless, some progress has been made in showing that BDNF has trophic effects in normal development (Kalcheim et al., 1987; Hofer and Barde, 1988). In contrast, the chance discovery of an extremely rich source of NGF—the salivary glands of male mice—and the early availability of an effective NGF antiserum allowed relatively rapid progress in understanding the biology and chemistry of NGF at a time when the techniques for accomplishing this sort of work were rudimentary by today's standards.

In spite of the power of modern molecular biology, progress in discovering and sorting out the physiology of additional trophic molecules (which, like NGF, probably exist in extremely small quantities in neural targets) may be slow—similar, perhaps, to the gradual discovery of additional neurotransmitters that followed the identification of acetylcholine and norepinephrine in the early part of the century. By the same token, differences among the actions of various trophic agents are likely to be just as remarkable as the differences that have been found among the actions of various neurotransmitters.

Uncertainties about the Biology of NGF

The integrity of the evidence supporting the view that NGF is a trophic agent for neuronal survival and neuritic growth among dorsal root and sympathetic ganglion cells tends to obscure some persistent problems. Pre-eminent among these is the ultimate physiological source of this trophic agent. NGF is synthesized constitutively by target cells; that is, it is continuously produced and secreted, rather than stored and released on demand (Heumann et al., 1984; Shelton and Reichardt, 1984). Because the amount of messenger RNA for NGF in each target cell is exceedingly low, the consensus has been that the limited amount of NGF in the relevant tissues explains the competitive target-dependence of NGF-sensitive neurons. The targets of sympathetic and sensory neurons, however, are not homogeneous in their cellular composition. For instance, when sympathetic targets like the iris are maintained for several days *in vitro,* the level of NGF production by the tissue increases over time (Ebendal et al., 1980; Heumann and Thoenen, 1986). Even though NGF is unequivocally synthesized by smooth muscle and other target cells of sympathetic and sensory axons (Bandtlow et al., 1987), immunohistochemical staining with anti-NGF indicates that much of this newly made NGF is actually in glial (Schwann) cells (Rush, 1984; Finn et al., 1986; Wilson et al., 1986). Similarly, in the skin, which is a major target for dorsal root ganglion cells, both the cutaneous epithelium and the underlying mesenchyme synthesize NGF (although most is in the presumptive epidermis—Davies et al., 1987). Finally, sensory ganglion cells, which have a bipolar axon that innervates a peripheral *and* a central target, are now known to receive trophic support from *both* the peripheral target and the spinal cord (Carmel and Stein, 1969; Yip and Johnson, 1984; Johnson and Yip, 1985; Davies et al., 1986b; E. M. Johnson et al., 1986; Thoenen et al., 1987). This fact, coupled with the possibility of different or multiple trophic agents operating at each target, implies that a detailed explanation of the trophic dependence of a given neuronal class may be quite complex, not unlike the sensitivity of a neuronal class to various neurotransmitters and neuromodulators. Thus, although the presence of messenger RNA for NGF in sympathetic and sensory targets has confirmed the general notion of the target-derived trophic support, the detailed nature of this support remains to be elucidated and may be quite complex.

The regulation of NGF production, as well as its source, may also be more complicated than initially imagined. For example, glial cells associated with sympathetic axons, called Schwann cells, synthesize NGF and NGF receptors after nerve injury (Korsching et al., 1986; Taniuchi et al., 1986, 1988; Heumann et al., 1987; Johnson et al., 1988). In the absence of neuronal contact either early in development or after a nerve injury, glial production of NGF and NGF receptor is high, falling only when the normal association of axons and glial cells is established. This observation, together with what is now known about the effects of activity on trophic interactions (see Chapter 8) indicates that the regulation of trophic feedback may also be rather labyrinthine.

The fact that NGF can be produced by Schwann cells and that this synthesis is influenced by the presence of axons raises the further possibility that neurons may normally acquire a portion of the trophic support they need over the whole traverse of the axonal projection to the periphery, rather than exclusively at the level of the target. Such an arrangement could explain the long-standing observation that the survival of neurons whose axons have been cut distally is greater than the survival of neurons whose axons have been cut close to the parent cell body (Lieberman, 1971). This result after axotomy has sometimes been ascribed to the greater "injury" to the neuron from a proximal lesion. An alternative explanation is that a longer axon "stump" increases the amount of trophic support that the injured cell can still acquire. Ongoing trophic support acquired by axons rather than by axon terminals might also help account for the prolonged survival of spinal motor neurons and ganglion cells following amputation or axon ligation in adult mammals (Chapter 6).

Another unresolved issue about the biology of trophic factors concerns the actual manner in which they generate competition among the axons that innervate a target. A conventional view has been that trophic agents are more or less freely diffusible; hence the ability of denervated tissues like the iris to induce sprouting among intact axon terminals that are at some distance from the trophic source. If trophic support were, as a general rule, acquired by the uptake of freely diffusible molecular signals, competition among axons might be based on local availability of the trophic agent in the vicinity of target cells. Consistent with this idea is the fact that in many autonomic and sensory targets, axon terminals ramify in only loose association with the target cells. In smooth muscle, for instance, the axon terminals are often several microns from the nearest smooth muscle or other cell. At many neural targets, however, closely apposed and highly specialized synapses are formed between the innervating axons and the target cells, as in the case of muscle and ganglion cell innervation. In these instances, in which trophic dependence of neurons on their targets is equally apparent, target-derived trophic support may be provided specifically at synaptic sites. Furthermore, trophic molecules may well be bound to the highly specialized extracellular matrix which occurs at some such sites (Sanes and Hall, 1979; Sanes et al., 1980; Sanes, 1983). Were trophic support to be specifically provided at synaptic sites by *nondiffusible* molecular signals, the arena for competition would be limited to those axons innervating the same target cell.

The Significance of Trophic Molecules

Trophic molecules are obviously not the sole, or even the primary, determinants of neural connectivity. Many other classes of molecules are now known to be involved in the formation and maintenance of synapses. These include the molecular signals that mediate qualitative recognition of appropriate pre- and postsynaptic partners, such as the molecular agents of chemoaffinity (Chapter 1); molecular signals derived from presynaptic elements that regulate postsynaptic specializations, such as the clustering of receptors and other synapse-specific molecules (Fischbach et al., 1979; Usdin and Fischbach, 1986; Schuetze and Role, 1987; Smith et al., 1987); molecular signals arising from postsynaptic elements that induce presynaptic differentiation, such as the formation of active zones (Sanes et al., 1978, 1980); and molecular signals that mediate cell adhesion, a phenomenon that is also likely to be important in synaptogenesis and terminal growth (Edelman, 1983, 1986; Sanes and Covault, 1985). In none of these categories is the biology of the molecular signal well understood. It is therefore possible that any or all of these classes of molecules influence the growth and retraction of neurites. The cell adhesion molecule N-CAM, for example, is expressed only at the neuromuscular junction in adult muscle but is present extrasynaptically during development and following denervation (Covault and Sanes, 1986). This fact, together with evidence of an inhibitory effect of N-CAM antibodies on axon terminal growth, raises the possibility that this and other cell adhesion molecules are important in phenomena like sprouting at the neuromuscular junction (Booth and Brown, 1987; Booth et al., 1987).

There is, however, a fundamental difference between trophic molecules like NGF and other molecules relevant to synaptogenesis. Trophic agents are defined by their regulatory effects on neuronal survival and growth. In contrast, other molecules involved in synaptogenesis are primarily concerned not with feedback and neural adjustment but with the process of making a functional synapse. In this sense, trophic mechanisms operate in parallel with synaptogenic mechanisms but for a broadly different purpose: the establishment of appropriate numbers of neurons, neuronal branches, and synaptic connections rather than the construction of synapses *per se*.

The overriding importance of information about NGF and other trophic molecules for the present argument is the answers these observations provide to several specific questions about the nature of trophic relationships. In particular, what fuels the evident competition among nerve cells for survival and for the deployment of innervation? And what links the size and form of the body to the organization of the neural pathways that serve it? These questions will be no less compelling if the particular paradigm suggested by the present understanding of NGF ultimately proves misleading. What is now known about the biology of NGF, however, strengthens the overall argument of the trophic theory by providing an example of how a target-derived signal can, by limited availability, generate competition among the innervating nerve cells both for survival and for axonal and dendritic growth. In this way, trophic agents appear to establish a molecular linkage between the body and the concatenation of neural connections interposed between primary motor and sensory neurons and progressively more remote neural centers.

EIGHT

Effects of Neural Activity on Target Cells and Their Trophic Properties

THE MAJOR business of the nervous system—perception of the body's internal and external circumstances and coordination of the appropriate responses—is mediated by the electrical activity of neurons. Activity, therefore, is likely to affect many neuronal properties. Most obviously, the electrical activity of nerve cells modulates their secretory behavior, the process on which chemical synaptic transmission is based. Furthermore, neural activity has a profound effect on the metabolism of excitable cells, which is largely devoted to maintaining appropriate intracellular concentrations of ions important for electrical signaling in the face of unfavorable concentration gradients. The level and pattern of electrical activity among nerve cells—that is, the number and arrangement of action potentials that pre- and postsynaptic cells manifest per unit time—also influence trophic interactions. As a result, neural activity can alter the connections between nerve cells and their targets.

Effects of Neural Activity on Target Cells

The most thoroughly studied effects of neural activity on the properties of target cells are those on the innervation of vertebrate skeletal muscle (Harris, 1974; Purves, 1976b; Grinnell and Herrera, 1981; Purves and Lichtman, 1985b). When the innervation of a muscle is removed by cutting the relevant motor nerve, a series of remarkable changes occurs. Within a few days, the denervated muscle fibers show a marked alteration in the passive and active electrical properties of their plasma membranes, become supersensitive to the action of acetylcholine (the normal transmitter at the vertebrate neuromuscular junction), and begin to lose myofibrils (a phenomenon that is ultimately apparent as grossly visible atrophy of the denervated muscle). Moreover, muscle fibers—which are normally resistant to further innervation, as by an implanted foreign nerve—become receptive to innervation following denervation and serve as a potent stimulus to axonal sprouting.

In the 1960s, when many of these phenomena were first described, the responses of muscle fibers to denervation were usually explained in terms of the loss of a hypothetical chemical signal continuously secreted by motor nerve terminals. Because a number of experiments provided circumstantial support for this interpretation (e.g. Miledi, 1960a,b), the possibility that electrical activity might be the agent of these effects was not thoroughly tested. In the early 1970s, however, several investigators belatedly asked whether the effects of denervation could be elicited simply by paralyzing an innervated muscle and, conversely, whether denervation changes could be forestalled by direct electrical stimulation of muscle. Paralysis was achieved in these experiments by blocking the nerve for several days with an implant containing a local anesthetic; direct stimulation was achieved with implanted electrodes. The results were unequivocal: many of the effects of denervation were brought on by the paralysis of a normal muscle, and with minor qualifications, all of them were prevented by direct stimulation of a denervated muscle (Jones and Vrbová, 1970; Lømo and Rosenthal, 1972; Lømo and Westgaard, 1975, 1976). Although none of these experiments ruled out an activity-independent chemical signal operating between nerve and muscle, they showed unambiguously that the level of activity of a muscle fiber serves to regulate a great many of its properties.

In general, the absence of activity affects neurons in much the same way that denervation or paralysis induced by local anesthesia affects the properties of muscle fibers. Thus, reducing activity by various means in autonomic ganglion cells causes decreased protein synthesis, similar to atrophic changes in denervated muscle (Black, 1978); increased sensitivity of the neurons to acetylcholine, the normal ganglionic transmitter (Kuffler et al., 1971; Roper, 1976; Dennis and Sargent, 1979; but see Dunn and Marshall, 1985); generation of a stimulus that elicits renewed growth (sprouting) of nearby axon terminals; and renewed receptiveness to innervation (Roper and Ko, 1978; Dennis and Sargent, 1979; Johnson, 1988). In one way, however, the effect of denervation on neurons is less drastic than the effect on muscle fibers. Chronically denervated sympathetic ganglion cells continue to function indefinitely (Voyvodic, 1987b), whereas chronically denervated skeletal muscle fibers atrophy progressively over a period of months and are eventually replaced by scar tissue (Tower, 1939). This difference is perhaps explained by the fact that denervated ganglion cells and other neurons continue to derive trophic support from the targets they innervate, whereas terminal targets like muscle fibers do not have this resource.

Work carried out in the vertebrate central nervous system, for the most part using anatomical measures, indicates that the properties of central neurons are also affected by neural activity. In the vestibular and auditory nuclei (Levi-Montalcini, 1949; Powell and Erulkar, 1962; Smith et al., 1983; Born and Rubel, 1985), the olfactory system (Matthews and Powell, 1962), the pontine and inferior olivary nuclei (Torvik, 1956), the lateral geniculate nucleus (Cook et al., 1951; Wiesel and Hubel, 1963a; Guillery, 1973a,b), the somatosensory cortex (Woolsey and Van der Loos, 1970; Van der Loos and Woolsey, 1973), the visual cortex (Rakic and Williams, 1986), and the lateral motor columns of the spinal cord (Okado and Oppenheim, 1984), the removal of innervation early in development causes a loss of the target neurons or atrophic changes in the surviving cells (Figure 8.1). The effect of reduced activity has been distinguished from denervation in such cases as the loss of dendrites from cells in the nucleus laminaris of the chick auditory system after destroying one ear (Smith et al., 1983; Deitch and Rubel, 1984) and the atrophy of neurons in the lateral geniculate nucleus of the cat after lid suture (Wiesel and Hubel, 1963a). The mechanism of postsynaptic atrophy after the removal of afferents is not known. These effects are nevertheless quite similar to the atrophic changes that follow denervation, or simply reduction of activity, in muscle or ganglia.

The gist of these observations is that silencing a neural pathway has profound effects on a wide spectrum of target cell properties. Some of these effects, such as the stimulation of presynaptic sprouting, are manifestations of trophic interactions between neurons and target cells. Such observations therefore imply that neural activity has the ability to affect trophic mechanisms.

Influence of Neural Activity on Retrograde Trophic Signaling

During the mid-1970s when the importance of neural activity in denervation effects was being clarified, attention was only beginning to be directed toward the possibility that targets such as muscle fibers and ganglion cells might be a source of trophic agents of the sort exemplified by NGF (e.g. Harris, 1974; Purves, 1975, 1977). As a result, the influence of neural activity on the trophic properties of target cells was not immediately tested. Within a few years, however, the issue of

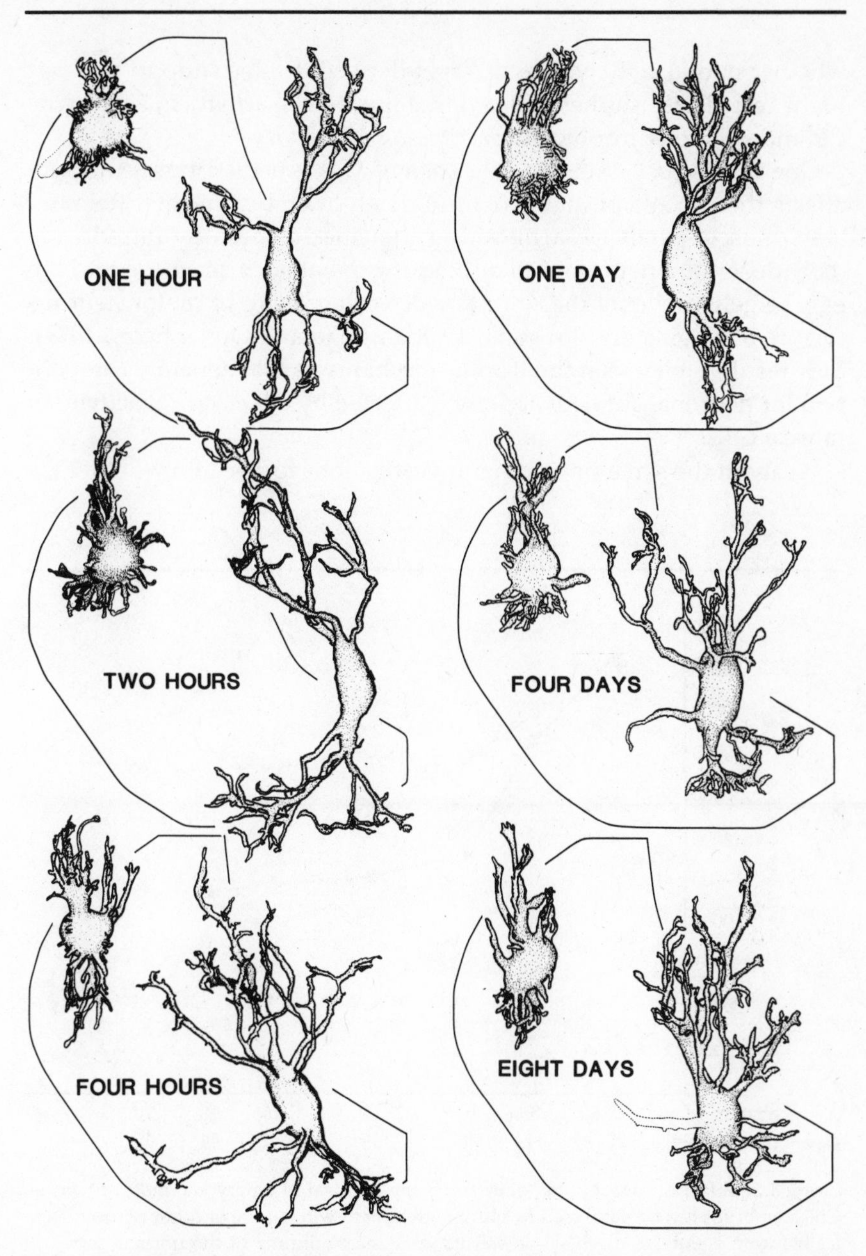

Figure 8.1. Effect of denervation on the morphology of some classes of neurons. Neurons from the chick nucleus laminaris, a third-order station in the auditory system, are shown at different intervals after selective removal (10 days after hatching) of innervation to the dendrites that extend downward. Beginning almost immediately and extending over several days, the denervated dendrites retract, whereas the innervated dendrites remain normal. This difference implies a local effect on the configuration of the postsynaptic neuron. The neurons are visualized by Golgi staining; the two cells shown at each stage indicate the range of dendritic geometries, which vary systematically as a function of neuronal position within the nucleus. The outline indicates the nuclear border. The same result is observed when activity in the pathway is reduced without denervation by destroying the ear. (From Deitch and Rubel, 1984)

whether neural activity affects target-derived trophic support was put to the test. These studies showed that the level of activity in a pathway can indeed affect trophic interactions.

One of the first issues to be explored was whether muscle activity affects the process of motor neuron death in embryonic life. Remarkably, muscle paralysis, achieved in the chick embryo by introducing daily doses of a neuromuscular blocking agent such as curare into the egg, largely prevents the normally occurring death of motor neurons (Figure 8.2; Laing and Prestige, 1978; Pittman and Oppenheim, 1979). This result implies that the trophic mechanisms that generate competition for neuronal survival are much affected by the electrical activity of muscle cells.

At about the same time, other investigations focused on whether the

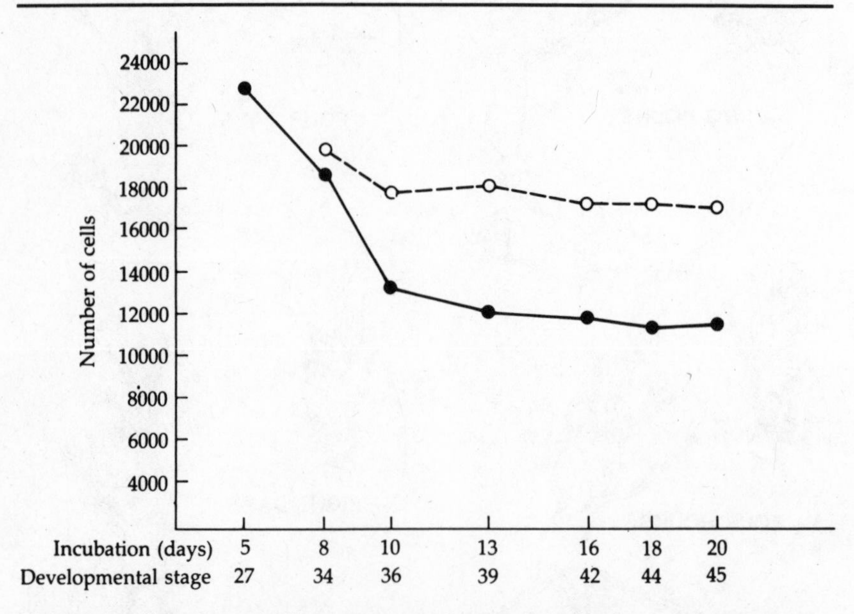

Figure 8.2. Effect of muscle fiber activity on the survival of embryonic motor neurons. Chick embryos can be paralyzed by chronic treatment with curare or other neuromuscular blocking agents. If the block is started near the beginning of the normal period of spinal motor neuron death (about the fifth day of incubation; see Chapter 3), the degeneration of these neurons is largely prevented. The solid line shows the normal course of motor neuron loss; the broken line shows the reduced loss during paralysis. When the neuromuscular block is removed by stopping the treatments, the motor neurons that were spared go on to die (not shown). Each point represents the average of cell counts in several different embryos. (After Pittman and Oppenheim, 1979)

neonatal rearrangement of mammalian neuromuscular connections (Chapter 5) is sensitive to activity. In these experiments, the rate of input elimination was assessed as a function of increased or decreased muscle activity. Recall that most skeletal muscle fibers in mammals and other vertebrates are initially innervated by several different axons; polyneuronal innervation is subsequently reduced to the adult one-on-one relationship over several weeks in early postnatal life. When neuromuscular activity is decreased by chronic paralysis, produced by local anesthesia or some other means, muscle fibers tend to retain innervation from multiple axons for a longer period (Benoit and Changeux, 1975, 1978; Riley, 1978; Thompson et al., 1979; Brown et al., 1982; Caldwell and Ridge, 1983; Bourgeois et al., 1986). A reduction of activity in peripheral pathways also slows the rate of input elimination among developing autonomic ganglionic cells (Jackson, 1983). Conversely, when the level of neuromuscular activity is increased to higher than normal rates by chronic electrical stimulation, muscle fibers lose inputs more rapidly than under normal conditions (O'Brien et al., 1978; Srihari and Vrbová, 1978; Thompson, 1983; Bourgeois et al., 1986). In accord with these results, vertebrate muscle fibers that *normally* lack action potentials—namely slow, or tonic, muscle fibers—retain multiple innervation indefinitely (Lichtman et al., 1985).

Another issue concerning the influence of electrical activity on trophic interactions is whether activity affects the *maintenance* of neural connections. When muscle activity is blocked by chronic application of a local anesthetic or a neuromuscular blocking agent in an adult rodent, motor nerve endings on the inactive muscle fibers begin to sprout (Figure 8.3; Brown and Ironton, 1977; Pestronk and Drachman, 1978; Holland and Brown, 1980; Pestronk and Drachman, 1985; Yee and Pestronk, 1987). Similar experiments in an autonomic ganglion show that reducing activity by the application of a local anesthetic to the preganglionic nerve causes sprouting. Thus, blocking action potential conduction for several days in the cervical sympathetic trunk of the guinea pig increases the size of synaptic potentials recorded in superior cervical ganglion cells, a result again indicating the sprouting of terminals on the surfaces of inactive target cells (Gallego and Geijo, 1987).

A particularly attractive site for exploring the effects of activity on trophic action is the innervation of smooth muscle by sympathetic ganglion cells, since the agent of trophic support in this instance is known to be nerve growth factor. In fact, denervated smooth muscle, like denervated skeletal muscle, presents a powerful stimulus to nerve terminal growth (Olson and Malmfors, 1970). Moreover, denervated

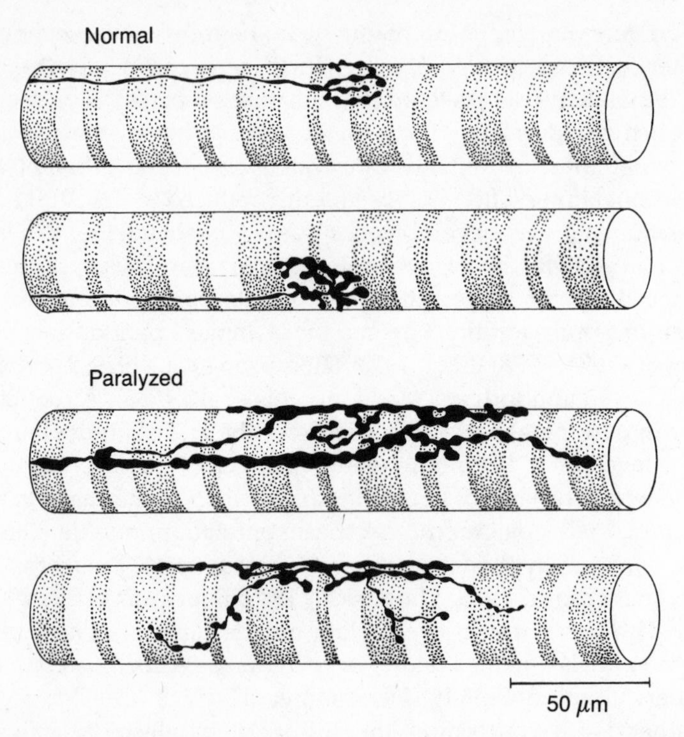

Figure 8.3. Renewed growth of motor nerve terminals induced by muscle paralysis in adult mice. (*A*) Normal terminal arbors at endplates in a soleus muscle, stained by a conventional histological method (zinc iodide-osmium treatment). (*B*) Terminal arbors on fibers from a muscle treated for 7 days with daily local injections of a neuromuscular blocking agent. The presynaptic terminals show extensive growth, though it is largely restricted to the surface of the target cell. (After Holland and Brown, 1980)

smooth muscle, when maintained in culture for several days shows increased levels of NGF synthesis (Heumann and Thoenen, 1986). The fact that under these circumstances the muscle is paralyzed implies an effect of inactivity on the availability of NGF. This interpretation has been complicated, however, by the observation that messenger RNA for NGF does not increase when the iris is denervated *in situ* (Shelton and Reichardt, 1986).

Taken together, these experiments indicate that the target-dependent death of innervating neurons, the normal rearrangement of connections in development, and the maintenance of adult connections are all affected by the level of electrical activity in the target cells. Thus, the regulatory signals for neuron survival and terminal growth, whatever they may be, are themselves susceptible to the influence of neural activity.

Influence of Neural Activity on the Arrangement of Competing Inputs

A less direct sort of evidence about the role of activity in the normal establishment of neural connections has come from observations of the deployment of synapses in autonomic ganglia. This work concerns the way in which *sets* of terminals arising from different axons are distributed among the target neurons. The ultimate arrangement of preganglionic innervation observed in mature ganglia is difficult to explain without recourse to the effects of neural activity on trophic feedback.

In parasympathetic ganglia, in which neurons are often innervated in maturity by a single preganglionic axon, anatomical and electrophysiological assessments indicate that the target cells captured by a particular axon (the neural unit) are widely scattered (Figure 8.4) (Lichtman, 1980; Hume and Purves, 1983; Purves and Wigston, 1983; Forehand and Purves, 1984). A similar dispersion of the target cells innervated by an individual motor neuron (the motor unit) is apparent in skeletal muscle (Burke, 1981). In spite of their wide distribution, most target cells in the neural unit are, in maturity, heavily innervated by the axon that supplies them. Since all of the target cells were, at an earlier time, innervated by several axons that each elaborated relatively few synaptic boutons, this final arrangement indicates that as development proceeds, each axon tends to make more and more synapses on fewer and fewer target cells. Because of the functional similarity between the muscle fiber endplate and the surface of a ganglion cell body (see Figure 5.4), the same reasoning applies to the innervation of skeletal muscle.

Evidently, once an axon makes *some* synapses on a target cell, the axon is informed that the target cell in question is potentially a favored site for the elaboration of additional synapses. In much the same way, the synaptic terminals that will ultimately be lost from each target neuron must, as a set, be informed that the neuron is one on which they will not flourish. This feedback cannot arise simply from the concentration of trophic signals in the vicinity of the postsynaptic cell, since a gradient of trophic molecules would not explain why adendritic target neurons, or skeletal muscle fibers, are usually innervated by *one* axon to the apparent exclusion of all others (Figure 8.4). If feedback depended simply on the local concentration of a trophic signal, there

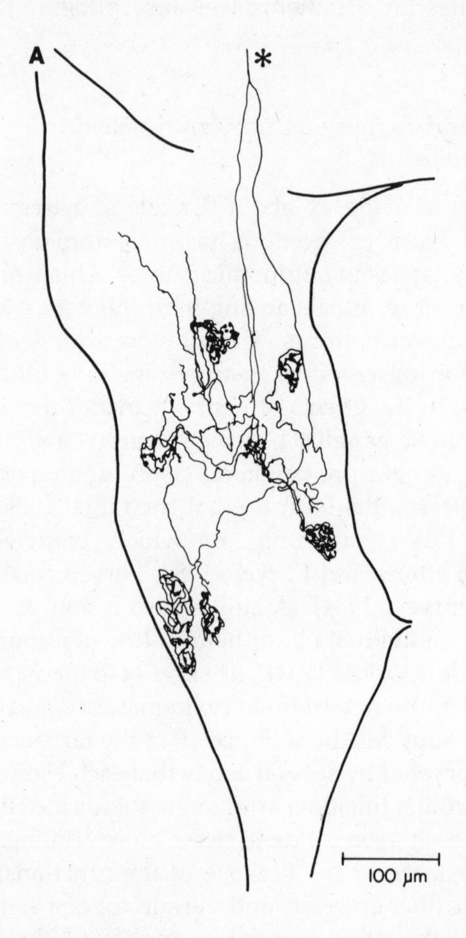

Figure 8.4. Avidity of an innervating axon for a few distributed target neurons within an autonomic ganglion. (*A*) Arborization of a single labeled axon in a rabbit ciliary ganglion. The outline shows the size and shape of the ganglion; the asterisk indicates the site at which the axon was injected with an intracellular marker. Although the axon ramifies widely, it elaborates many synapses on a few neurons (indicated by clusters of synaptic boutons) scattered among several hundred potential target cells. (*B*) Portion of the terminal arborization of a single labeled axon in another ganglion, showing synaptic boutons densely concentrated on three target neurons. The field contains dozens of similar target cells, which are not apparent because they are not outlined by terminals from the labeled axon. The asterisk indicates the position of one such "overlooked" target cell, sandwiched between two heavily innervated neurons. The impression that the labeled terminals are sharply confined to a few densely innervated cells has been confirmed by electron microscopy. (*A* from Hume and Purves, 1983; *B* from Forehand and Purves, 1984)

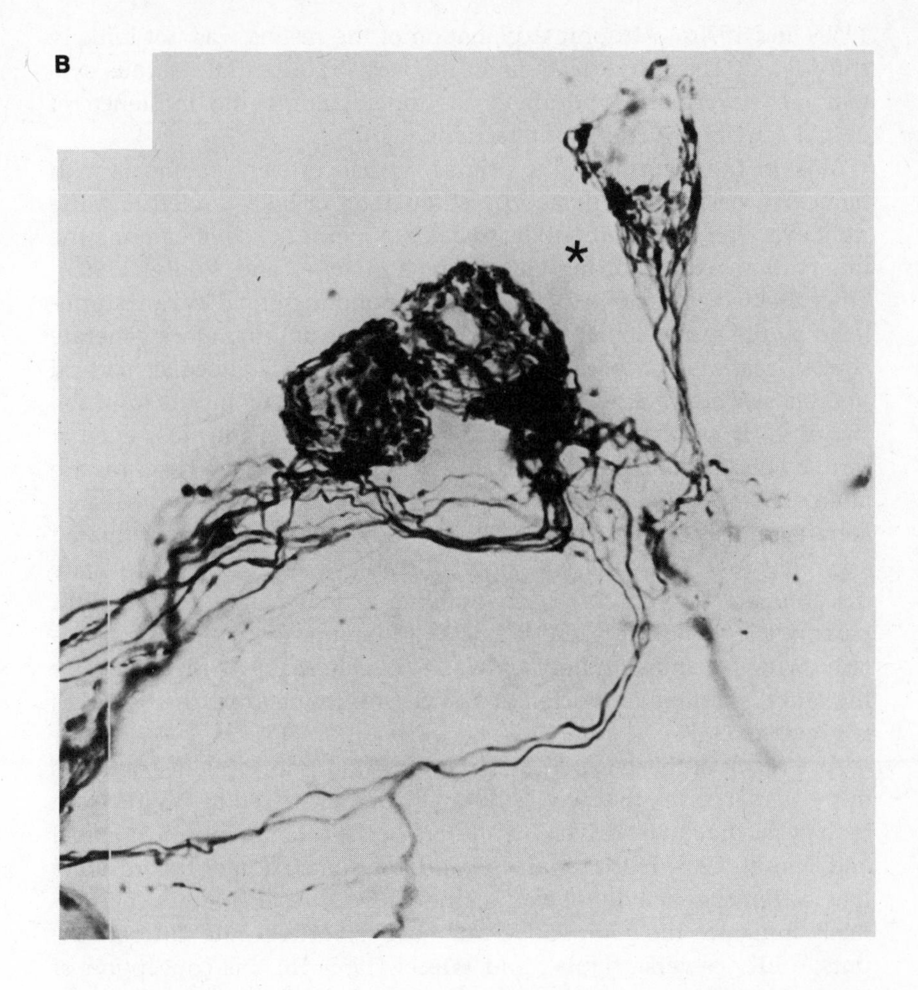

would be no bias toward innervation by any particular set of presynaptic terminals. To explain the way these anatomically simple target cells are actually innervated requires a mechanism that somehow allows *all* the terminals on a particular target cell to compete as a set during the formation and maintenance of connections.

Influence of Neural Activity on Convergent Innervation

Yet another indication of the influence of activity on neural connections has come from studies of convergent innervation in the mammalian visual system. Because much of this work was carried out in the

1960s and 1970s, a trophic explanation of the results was not initially considered. These observations in the vertebrate central nervous system are nevertheless pertinent to understanding the influence of neural activity on trophic interactions.

Most of the neurons in the primary visual cortex of mammals with binocular vision are activated by stimulation of both the left and the right eye. This arrangement depends on patterns of activity among the innervating axons during development (Hubel and Wiesel, 1965). Since the two eyes move together, corresponding retinal loci are stimulated by the same elements of the visual scene and therefore generate similar patterns of neural activity. As a result, binocular cortical neurons are activated more or less simultaneously by inputs from the two eyes. If, on the contrary, a cat or a monkey is made cross-eyed at birth by cutting some of the extraocular muscles, or if the two eyes are alternately occluded, cortical neurons that would normally be activated simultaneously by inputs related to the right and left eyes are activated *asynchronously*. This difference occurs because, when the eyes are made disconjugate by surgery, corresponding retinal points see different parts of the scene and therefore have different activity patterns. Similarly, when an animal is only allowed to see alternately with one eye or the other, simultaneous activation of cortical inputs from the two eyes is prevented.

As a result of these experimental interventions in early life, neurons in the visual cortex that would normally be innervated by inputs from both eyes come to be driven by one eye or the other (Figure 8.5; Hubel and Wiesel, 1965; Hubel et al., 1977). One explanation of this result is that *synchronous* activity of axons innervating the same target cell normally mitigates the competitive interactions between the different inputs, and vice-versa (Hubel and Wiesel, 1965). In this conception of things, the shift from predominantly binocular to predominantly monocular innervation of cortical neurons during rearing with a surgically created squint, or alternating eye occlusion, occurs because few, if any, neurons are synchronously activated by the two eyes under these circumstances.

The influence of activity on inputs to the same neurons in the visual system implies that such effects should be evident in other sensory systems in which the target neurons are innervated by simultaneously active axons. A case in point is the auditory system of the barn owl (Knudsen et al., 1977; Knudsen and Konishi, 1978a,b; Knudsen et al., 1982; Knudsen, 1985a,b). The owl normally uses binaural cues, that is differences in the timing and intensity of sound in each ear, to localize its prey at night (see Chapter 2). If one ear in young birds is experimentally occluded by inserting a rubber plug, the activity of inputs associated with the deafened ear is compromised, whereas the activity of inputs associated with the other ear is normal (Knudsen, Esterly, and Knudsen, 1984; Knudsen, Knudsen, and Esterly, 1984). After rearing under these conditions, the animal's ability to localize sound can be tested by comparing head movements in response to both auditory and visual stimuli. Normally, an owl responds to an auditory stimulus at a particular location in space by accurately shifting its gaze to that place, a phenomenon that indicates a high degree of registration between these two sensory systems. Discrepancies between visual and auditory responses thus serve as a measure of the effects of abnormal rearing. Owls raised to maturity with monaural deprivation behaved normally with the plug in place and were able to shift their gaze to a source of sound quite accurately. This result shows that connections in the auditory and visual systems adjust during development to an experimentally altered aural environment. However, when the plug was *removed,* the owls showed systematic errors in sound localization for which they could not compensate subsequently. As in the mammalian visual system, then, appropriate patterns of activity in the auditory pathway are necessary for the establishment of normal connections and for the accurate registration of two different modalities like audition and vision (Knudsen, 1985a).

The importance of modulating cortical connections according to the temporal pattern of activity among inputs to the same target cells is not hard to imagine. In the visual system, mitigation of competition among axons innervating the same neuron in the maturing cortex by simultaneous activity would ensure that only inputs activated by approximately the same part of the visual scene persist in innervating a particular cortical cell. Accordingly, the precise binocular innervation of cortical neurons needed for stereopsis, for example, can occur by obedience to a simple epigenetic rule. Furthermore, the ability of connections in the visual system to be modified according to their individual functions allows the neural map of the visual world to adjust to changes in the size and position of the eyes imposed by growth (Chapter 2). A similar argument can be made for the effects of activity in the developing auditory system.

A Model of Activity-Dependent Trophic Support

These observations make several points. First, the electrical activity of target cells influences their trophic properties. Second, the formation and maintenance of innervation among target cells proceed according

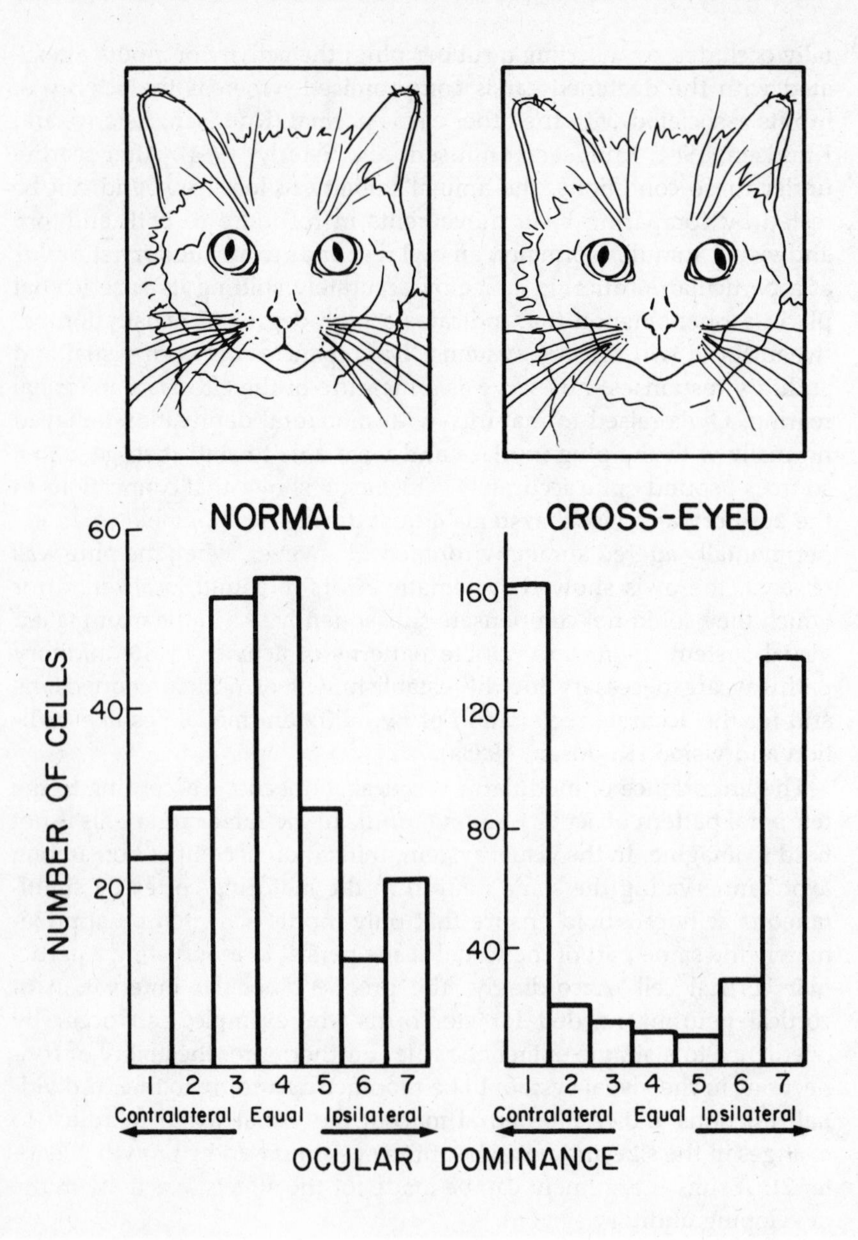
NORMAL
CROSS-EYED
NUMBER OF CELLS
60
40
20
160
120
80
40
1
2
3
4
5
6
7
Contralateral
Equal
Ipsilateral
OCULAR DOMINANCE

to *sets* of terminals defined by a common parent neuron and residence on a particular target cell. Third, competition between terminal sets for continued residence on a target cell is affected by the temporal ordering of activity among the competitors. Is there an explanation of the effects of activity on innervation that can rationalize these several facts?

The observed behavior of nerve terminals in muscle, ganglia, and central sensory systems could be explained if the acquisition of trophic support from target cells depends upon simultaneous neural activity. The history of this idea goes back a long way. The psychologist D. O. Hebb suggested in 1949 that *conjoint* activity of pre- and postsynaptic elements alters the efficacy of a synapse. For Hebb, this notion provided a theoretical explanation for the strengthening of connections that he presumed to occur during learning. As a result, connections thought to be strengthened by simultaneous activity are often referred to as "Hebbian synapses." Although there was little evidence on the issue in his day, Hebb saw the strengthening of connections by conjoint activity as a logical requirement of any neurophysiological explanation of learning. He also favored a structural explanation for the effect that he postulated: "When one cell repeatedly assists in firing another, the axon of the first cell develops synaptic knobs (or enlarges them if they already exist) in contact with the . . . second cell" (Hebb, 1949, p. 63). Ironically, the notion of synaptic strengthening by conjoint activity, for which Hebb is best remembered, was a relatively

Figure 8.5. Effects of temporal ordering of input activity on the innervation of neurons in the mammalian visual cortex. Animals with eyes in the front of the head, as distinct from wall-eyed animals, have binocular vision (both eyes seeing the same scene). As a result, corresponding points on the two retinas are subjected to roughly the same spatiotemporal pattern of visual stimuli. This situation can be changed by cutting some of the extraocular muscles, so that the two eyes—now with different directions of gaze—see different scenes. The effects of this procedure can be assessed in the cat by recording the responses of individual neurons in the visual cortex to stimulation of one eye or the other. The degree of cortical binocularity is shown by the ocular dominance histograms. Bars indicate different degrees of dominance by one eye or the other; 1 indicates neurons that respond solely to stimulation of the contralateral eye; 7 indicates neurons that respond only to stimulation of the ipsilateral eye; and 4 indicates neurons that are driven about equally by either eye. Normally, about 80 percent of neurons in the primary visual cortex of cats respond to stimulation of both eyes. In cats made cross-eyed at birth by cutting some of the extraocular muscles, only about 20 percent of the neurons are binocularly driven. The effect of being cross-eyed is to desynchronize the activity of the inputs arising from the two eyes. Asynchronous activation of axons innervating the same cortical nerve cell evidently foments competition between the convergent axon terminals. (After Hubel and Wiesel, 1965)

minor point in his theory about how mental constructs are represented by groups of self-exciting neurons, and he apparently considered this idea to be derivative (Milner, 1986).

In the several decades since Hebb's book was published, a number of neurobiologists have tried to revitalize Hebb's idea by suggesting a more detailed mechanism that might lead to strengthened connections as a result of conjoint activity (Stent, 1973; Lichtman and Purves, 1981; Fraser and Poo, 1982; Fraser, 1985). In G. Stent's model, the catalyst for a line of thought that began in the early 1970s, the proposed effect of conjoint activity is to moderate a voltage-dependent elimination of postsynaptic receptors. Such a sparing of receptors was postulated to occur because ligand-activated channels at a synapse tend to hold the subsynaptic membrane at a potential, called the reversal potential, which is less positive than the voltage attained extrasynaptically at the peak of the action potential. Accordingly, the receptors underlying any two terminals that activate the postsynaptic cell simultaneously tend to be protected from the presumed effects of postsynaptic action potentials, whereas the receptors at a synapse that fires at a different time from the majority tend to be destroyed. Because the effectiveness of a synapse depends on the presence of postsynaptic receptors, this effect would tend to strengthen some synapses and weaken others.

Another model is based on electrophoretic movement of postsynaptic receptors (Fraser and Poo, 1982; Fraser, 1985). The electrical activity of a postsynaptic cell is known to affect the distribution of receptors in its membrane by electrophoretic migration. This model proposes that the input which is most effective in depolarizing a target cell (that is, which innervates it most strongly) has the largest time-averaged electrical field. Such an input is therefore most efficient in localizing receptors in its vicinity. With respect to the conjoint strengthening of two inputs as a result of their simultaneous activity, any two axons that activate the target cell at the same time partially cancel each other's field effect. Accordingly, simultaneously active inputs have an advantage in localizing receptors vis-à-vis other inputs that fire asynchronously.

A third model is based explicitly on the acquisition of trophic support from target cells (Lichtman and Purves, 1981; Purves and Lichtman, 1985b). This particular variation on Hebb's original theme was motivated by the increasing evidence that target cells provide specific trophic molecules to the neurons that innervate them (see Chapter 7). In this scheme (Figure 8.6), each terminal branch on a target neuron must acquire some minimum amount of trophic support

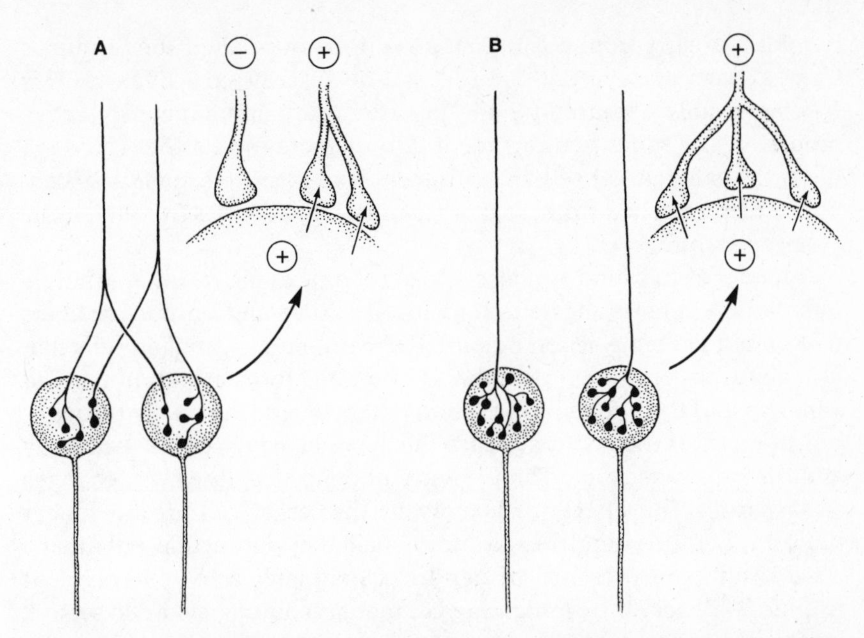

Figure 8.6. Possible linkage of trophic feedback to neural activity. (*A*) Initially, ganglion cells that lack dendrites are innervated by several axons. The assumption of the model depicted here is that acquisition of trophic support depends on simultaneous activity (depolarization) of the pre- and postsynaptic cells (blowup on right) and that trophic signals are not widely diffusible. The number of synaptic boutons initially made by each axon on a given target cell is stochastic. In consequence, whenever an axon with a plurality of terminals on a target cell is active (+), it gains trophic support at the expense of an inactive axon (−) that makes fewer synaptic contacts. Ultimately, the minority terminals are withdrawn from the target cell, and the "losing" axon elaborates additional terminals on other target cells where it had the initial advantage. However, if the two axons were consistently active together, they would not compete with one another. (*B*) The outcome of such activity-dependent trophic feedback among axons that are not synchronously active is to concentrate each axon's endings on a relatively few target cells.

in order to be maintained. To account for the phenomenon of competition among sets of synaptic endings, in contradistinction to competition among individual terminal elements, the acquisition of trophic support is taken to depend on the simultaneous activity of the pre- and postsynaptic elements; in this Hebbian scheme the agent of synaptic "strengthening" is a trophic molecule rather than the localization of receptors or some other effect. If the trophic support acquired by presynaptic endings depends on simultaneous pre- and postsynaptic depolarization, then individual postsynaptic cells can reward only *sets* of terminals arising from a competing axon, because all of the terminals from a particular parent neuron are necessarily co-activated. This scheme readily accounts for the persistence of simultaneously active inputs on the same neuron, for if two or more inputs that converge upon the same target cell are consistently co-activated, then they cannot be distinguished from a set of terminals that derives from the same parent neuron.

A model of this kind can also explain the puzzling fact that multiple innervation persists on neurons with dendrites and on muscle fibers that do not generate action potentials. In autonomic ganglion cells that lack dendrites, an action potential is generated by each suprathreshold stimulus and the effects of the action potential are felt equally throughout the part of the cell that bears the synaptic contacts (the cell body and the proximal axon). The presence of dendrites, however, changes this picture. The effects of activity in the target cell are no longer uniform, because dendrites do not usually support action potentials. As a result, the presence of dendrites may alter activity-dependent trophic feedback, thus mitigating competition among synaptic sets. A similar argument applies to the persistence of multiple innervation on tonic muscle fibers, in which the postsynaptic membrane is also incapable of supporting an action potential (Morgan and Proske, 1984; Lichtman et al., 1985).

Models in neurobiology should generally be regarded with suspicion. Consideration of these particular schemes, however, is justified by the fact that the way innervation is established and maintained in both the peripheral and the central nervous systems is unlikely to be understood without recourse to ideas of this general sort, in which synaptic strengthening is contingent upon an activity-dependent reward.

Some Complications

In addition to the difficulties posed by the incompleteness of evidence about the effects of activity on trophic interactions, particularly in the central nervous system, there are several other problems. One complication is the fact that the innervation of target cells is often inhibitory. In this instance, a presynaptic action potential does not depolarize the postsynaptic cell but simply increases its conductance to one or more ions that have equilibrium potentials at or near the resting potential of the postsynaptic cell. The general argument regarding modulation of trophic effects by depolarization does not apply in such cases. Whether removing or silencing the activity of inhibitory innervation affects the properties of target cells, trophic or otherwise, has not been explored. Nor is it clear that target-derived trophic signals affect inhibitory innervation, although this seems likely in view of the general purposes of trophic interaction.

Another problem concerns the sensory innervation of peripheral targets. Bipolar sensory neurons evidently receive trophic support from both their central and their peripheral projections (see Chapter 7). The central projections of such cells presumably acquire trophic support in much the same way as other central and peripheral projections discussed in the preceding sections. However, because the peripheral axons of these neurons do not establish synapses with their targets, a scheme based on the acquisition of trophic support by postsynaptic depolarization (see Figure 8.6) is obviously unsatisfactory. A similar problem is raised by the innervation of smooth muscles by autonomic ganglion cells, where the "synapses" comprise pre- and postsynaptic elements that are at some distance from one another, and where competition may therefore be based on local concentrations of trophic molecules rather than on a competition that unfolds on the surfaces of particular target cells. The acquisition of trophic support in these circumstances may also be activity dependent; but if so, a somewhat different model will have to be devised.

The Significance of Activity-Dependent Modulation of Neural Connections

The influence of electrical activity on the properties of postsynaptic cells adds another dimension to the regulation of neural connections. Broadly speaking, neural activity or its absence has two different, but related, effects on postsynaptic cells. A general effect of reduced activity on target cells, whether in the central or in the peripheral nervous system, is to cause varying degrees of atrophy and even cell death. From a functional perspective, this effect makes good sense, for a cell that has not succeeded in becoming innervated does not need to be metabolically active and may well be redundant. Conversely, uninnervated or inactive target cells apparently increase the availability of the trophic factors that stimulate the formation and maintenance of the innervation they normally receive. This additional effect of activity also makes good sense: a cell in receipt of insufficient synaptic contacts tries its best to attract innervation. The influence of activity on the trophic properties of target cells thus establishes a regulatory loop in which both neuron and target remain responsive to circumstances.

The apparent yoking of neural activity and retrograde trophic action also helps explain how information about the detailed function of competing axons can be imparted in the course of forming neural circuits. The phenomenon of chemoaffinity (see Chapter 1) is, by all evidence, the major basis of the qualitative accuracy of neural connections: intercellular recognition of surface labels allows pre- and postsynaptic elements to distinguish appropriate synaptic partnerships from inappropriate ones in the course of development and establishes appropriate topographic maps. There are, however, more subtle needs that could not be mediated by such labels. In the visual or auditory systems, for example, correct operation requires that particular target cells be innervated by inputs that have particular *functional* qualities (for instance, inputs that respond to the same part of the visual world or to a sound arising from a particular location). The discrimination of some of these functional attributes during the formation and maintenance of neural connections is evidently generated by the temporal ordering of activity among competing inputs. Thus, inputs with the same activity patterns tend to be retained on target cells, whereas inputs with different patterns tend to be rejected, in the sense that only one set or the other can remain in contact with the target cell. The influence of activity on the acquisition of trophic support provides a means by which such functional discriminations can be made. Adjustments of this type—that is, ones that take into account the function of the competing inputs—are useful not only in establishing the details of neural connections but also in allowing them to change progressively during growth or for some other reason.

In summary, the interplay of activity and trophic feedback can explain a number of otherwise puzzling observations. The common denominator of these observations is the innervation of target cells according to synaptic sets. Predicating trophic support upon the simultaneous activity of pre- and postsynaptic cells makes it possible to rationalize why sets of synapses derived from single axons, rather than single synapses, compete with each other and why inputs that share the same pattern of activity compete less strongly with one another than do inputs that have asynchronous patterns of activity.

NINE

Implications of the Trophic Theory of Neural Connections

THE EVIDENCE which supports the idea that neural connections in mammals and other vertebrates are importantly influenced by signals from neural targets, which in turn are subject to modulation by neural activity, has implications for several outstanding issues in contemporary neurobiology. These include the question of differences between neural mechanisms in vertebrates and invertebrates, the neuronal basis of learning and memory, and the notion that patterns of neural connections arise selectively during a Darwinian struggle in which only a fraction of the initial connections persists. Whereas recognition of the role of trophic interactions in the formation and maintenance of neural connections will not resolve any of these sometimes contentious issues, the trophic theory is pertinent to each of them.

Implications for Different Taxa

In general, biologists study a particular animal either because they find it intriguing in its own right or because they take it to be especially suited to solving a more general problem in which they are interested. Although both of these motives are fully represented within the ranks of modern neurobiologists, many of those who work on invertebrates do so because the relatively simple nervous systems of such animals allow a more direct approach to complex problems than is presently possible in vertebrates. However, the use of invertebrate neural systems as models for problems not easily addressed in vertebrates presents a dilemma: what if the nervous systems of very simple animals operate in a fundamentally different way from vertebrate nervous systems? Anxiety about this possibility has been made more acute by evidence that developmental strategies *are* different among taxa (see Chapters 3–4). This concern has sometimes led to a situation in which a suggestion that the nervous systems of simple animals might operate differently from those of vertebrates (e.g. Easter et al., 1985) is countered by the claim that invertebrate nervous systems are in all important respects similar to vertebrate systems, if only one looks at them in the right way (e.g. Murphey, 1986b). Although this controversy is likely to go on for some time, the issue of whether the trophic theory applies to the very large number of animals defined by the absence of backbones can be settled relatively easily.

The biological purposes of neural regulation by trophic signals in vertebrates are based on the need of the nervous system to remain informed about and responsive to changes in the body it must monitor and motivate. Among these purposes are the quantitative apportionment of innervation during development, the modulation of neural connections required by growth and changing form, and the facilitation of changes in neural organization that must occur as the form and function of species are modified during the course of evolution. Since *all* animals must solve these problems, trophic interactions—or their equivalent—should be widely evident. Certainly this logic applies to animals both with and without a backbone.

An example that helps underscore this point is the crustacean *Homarus americanus*. Lobsters of this species are fully formed at an early juvenile stage when the animal weighs less than 100 milligrams (Figure 9.1). However, these invertebrates, like many others, gradually grow to a large size; the largest lobsters on record weigh as much as 20 kilograms (Phillips et al., 1980). Thus, during a series of molts, a juvenile lobster may increase in bulk by a factor of 10^5 or more, a degree of growth which rivals that of the largest mammals. From the size and frequency of molting among smaller lobsters observed in the laboratory, it can be calculated that lobsters of 15 kilograms or more are at least a century old (C. K. Govind, personal communication). Such great and prolonged growth implies ongoing neural adjustment, just as in vertebrates.

The reorganization required by growth in the lobster could occur by the continual addition of neurons or, as in most parts of the vertebrate nervous system, by growth and rearrangement of axonal and dendritic branches. From what is known of the lobster nervous system, the means of adjustment appears to be the remodeling of existing neurons and their connections. Lobster limb muscles, for instance, are innervated by a small number of neurons (up to 8). To take a specific case, the closer muscle of the lobster claw is innervated by the same two excitor motor neurons in a juvenile animal as in an older animal whose claw is vastly larger (Govind, 1984). In short, the lobster, like verte-

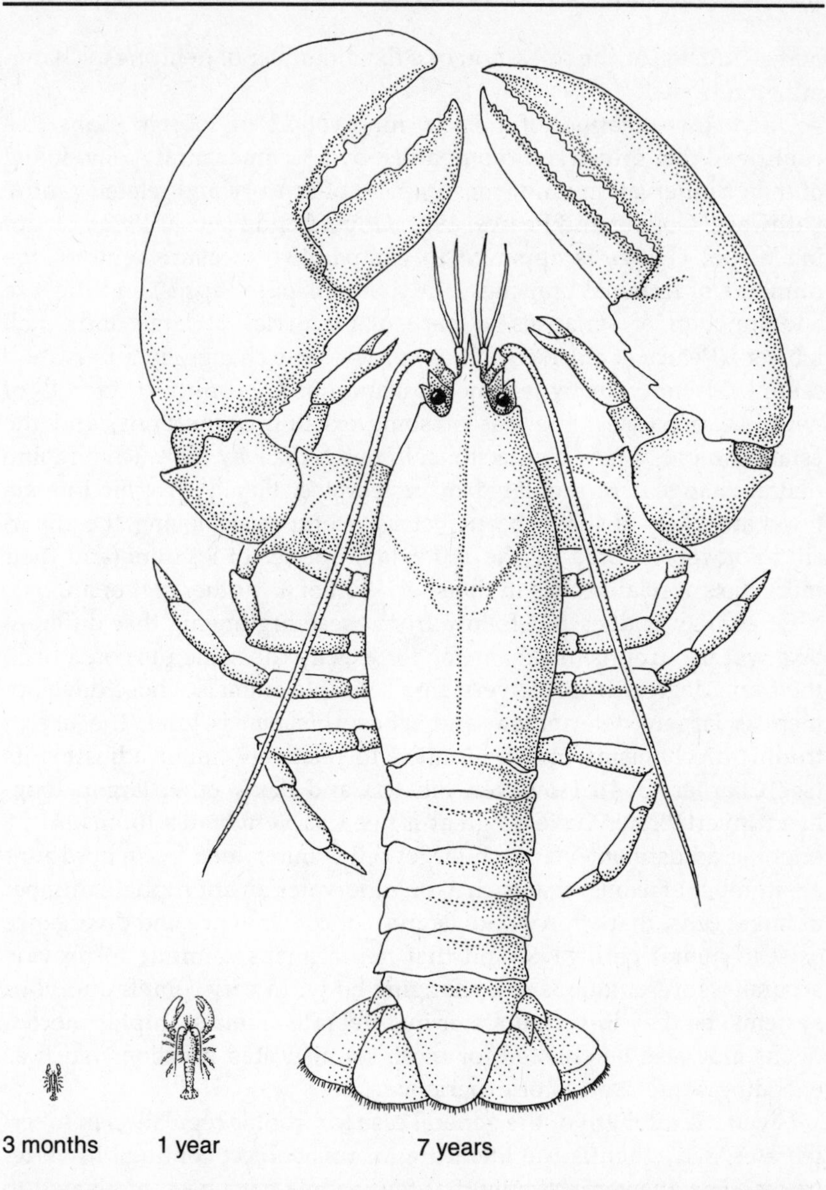

Figure 9.1. Growth of the lobster *Homarus americanus,* which may increase in weight by some 5 orders of magnitude during a lifetime that spans a century or more. As in vertebrates, a fixed number of nerve cells in the lobster must cope with a changing body. The growth of the lobster is also allometric. Lobsters begin life with claws that are symmetrical in form, but as the animal matures, one claw (here the left) becomes a crusher, and the other becomes a pincer. These obvious changes in size and form require neural adjustments by trophic interactions, just as they do in vertebrates. (After Govind, 1984)

brates, must gear the operation of a fixed number of neurons to changing targets.

Direct investigation of muscle innervation in lobster limbs has confirmed that growth is accompanied by a commensurate remodeling of motor innervation (and, presumably, of sensory and related central pathways; Govind and Pearce, 1981; Govind and Derosa, 1983). Ongoing neural change is apparent in comparative measurements of the numbers of terminal branches, the numbers of synapses, and the size and function of synapses in particular muscles of large and small lobsters (Pearce et al., 1985). These progressive changes in innervation seem likely to occur by trophic regulation of the continued growth of neuronal processes, the loss of some existing connections, and the establishment of new connections, in much the way that such ongoing neural changes are regulated in vertebrates. Finally, trophic interactions are likely to have been just as useful in achieving the diversification represented by the 160 extant species of lobsters (and their more distant relatives) as in the speciation of any other sort of animal.

Of course, it does not follow from these arguments that different taxa will use trophic mechanisms for exactly the same purposes or in the same degree. In some very small, simple animals whose development is largely determinate and whose lifespan is brief, the use of trophic mechanisms may be limited to relatively minor adjustments (see Chapters 3–4). However, lobsters and many other larger, long-lived invertebrates have as great a need as vertebrates to ensure by ongoing adjustment that each target cell is innervated by an appropriate number of axons, that each axon innervates an appropriate number of target cells, that appropriate degrees of convergence and divergence exist in neural pathways, and that neural maps continue to provide accurate representations of a changing body. In very complex nervous systems, be they in vertebrate or invertebrate animals, trophic mechanisms may also be enlisted for more sophisticated functions, such as encoding some aspects of experience.

Given the strength of the general case for trophic regulation in invertebrates, why then is the literature on this subject confined to vertebrates? The answer is simply that few people have been motivated to explore trophic interactions in invertebrates. For sound historical reasons, the focus of invertebrate neurobiology has been elsewhere. Consequently, few systematic studies have been conducted in invertebrates on questions such as the dependence of neurons on their targets for survival, the dependence of neuronal size and form on targets, input elimination and synaptic rearrangement during development, the dependence of convergence on neuronal form, the correlation of neuronal form and innervation during ontogeny, the correlation of the form and innervation of homologous neurons in related invertebrates of different size, the maintenance of neural connections in maturity through target interactions, the ongoing remodeling of neural connections in living animals, or the existence of target-derived growth factors. Such studies, each of which could be done in various invertebrates, would be rewarding not only for their contribution to comparative neurology but also for the further insight they would certainly provide into the cellular and molecular bases of trophic interactions.

Implications for Learning and Memory

The neuronal basis of learning and memory represents a second context in which the trophic theory may generate new perspectives. In general, studies aimed at understanding how animal behavior is modified by experience—that is, how animals learn—have centered on how neural activity affects circuits assumed to be anatomically stable. The reasons for this emphasis are again both historical and pragmatic. Historically, the traditional view has been that neural connections are largely hard-wired (see Chapter 1). "As far as I know," R. W. Sperry once remarked, "among all the synapses that have been observed in the history of neurology, not one of them has yet been demonstrated to have been implanted by learning" (1967, p. 142). On the pragmatic side, much behavioral change occurs so quickly that a connectional basis for it is unlikely in principle (but see Hebb, 1949; Crick, 1982). Nevertheless, for certain kinds of learning, namely changes in behavior that develop slowly and last a long time, encoding experience by altering the arrangement of neural connections is an altogether plausible strategy.

To understand more specifically how the trophic theory may be relevant to thinking about learning and memory it is useful to consider the present consensus about the cellular and molecular bases of these phenomena. The major lines of research during the last several decades into the neural correlates of behavioral change are easily traced. Beginning with the work of B. Katz and others in the 1950s, compelling evidence has accumulated to indicate that the frequency of action potentials in a neural pathway has a marked influence on the short-term efficacy of synaptic function (Del Castillo and Katz, 1954; Katz, 1969). If one nerve impulse follows another in rapid succession at the neuromuscular junction, the amplitude of the postsynaptic response elicited by the second action potential is increased for several hundred milliseconds. The effect, which arises presynaptically, is due to an increase in the number of transmitter quanta released by the second impulse of the pair (Katz, 1969; Zucker, 1982). This phenomenon is called facilitation. More intense activity at the neuromuscular junction—a train of impulses—leads to a form of facilitation called post-tetanic potentiation, which may last for minutes or longer (Hutter, 1952; Del Castillo and Katz, 1954; Liley, 1956). These effects at the neuromuscular junction appear to arise from differences in the influx (or buffering) of calcium ions, which normally triggers transmitter release from the presynaptic terminal (Katz and Miledi, 1968). This interpretation, called the residual calcium hypothesis, still seems the best explanation of the effects of activity on the efficacy of neuromuscular transmission (Zucker, 1982, 1987). Although these phenomena are most thoroughly understood at the neuromuscular junction, the effects of activity on synaptic efficacy are quite general. Thus, studies of mammalian autonomic ganglia (Larrabee and Bronk, 1947), the spinal cord (Lloyd, 1950), and the hippocampus (Bliss and Lømo, 1973; Swanson et al., 1982; Teyler and DiScenna, 1987), have revealed similar effects of prolonged stimulation on the efficacy of synaptic transmission. More recently, work on the mechanisms of synaptic potentiation in the mammalian hippocampus has shown that a long-lasting change in synaptic efficacy at this site can also be caused by a postsynaptic calcium flux, which is in turn dependent on simultaneous pre- and postsynaptic depolarization (Kelso et al., 1986; Teyler and DiScenna, 1987; Smith, 1987).

Intriguing though these electrophysiological studies in vertebrates may be, for biologists interested in understanding how experience generates altered responses, the vertebrate nervous system has an obvious drawback: the synaptic changes observed cannot be easily related to animal behavior. The gulf between synaptic change and behavior is hard to bridge in vertebrates simply because the anatomical complexity of the nervous system usually precludes direct examination of the neuronal elements underlying behavioral responses in intact animals. As a result, activity-induced changes at the neuromuscular junction or in the hippocampus have so far served as models of what *might* happen during learning rather than as demonstrations of what *does* happen when learning occurs. To overcome this difficulty, many neurobiologists turned their attention to simpler animals—various invertebrates—in which circuits mediating easily observed behaviors can be identified and examined directly by impaling the relevant neurons with a microelectrode.

The most successful implementation of this approach has explored the neural basis of learning in the marine mollusc *Aplysia californica* (Kandel, 1979a,b,c; Kandel and Schwartz, 1982). This work has made at least two major contributions. First has been to demonstrate that analyzing the neural basis of behavior from the cellular perspective is indeed feasible in relatively simple animals. Changes in the properties of identified nerve cells mediating a reflex response can in fact be correlated with behavioral change, as measured by the vigor of reflex activity (Castellucci et al., 1970, 1978; Castellucci and Kandel, 1974; Kandel, 1979a,b,c). The second contribution has been a molecular model, supported by considerable evidence, to explain how the neuronal changes associated with altered behavior may occur (Kandel and Schwartz, 1982; Goelet et al., 1986; Schwartz and Greenberg, 1987).

In this and other work taking a similar approach, the proposed explanation of learning in the short-term is based on modification of ion channels in the presynaptic terminal membrane (Krasne, 1976; Quinn, 1984; Carew and Sahley, 1986; Goelet et al., 1986). Such changes can arise from activity in the altered synapses themselves, as also occurs in simple vertebrate models such as the neuromuscular junction, or from the release of a neurotransmitter by another pathway onto the altered synapses. In either case, modified channels affect the electrical characteristics of the terminal and thus alter the amount of neurotransmitter released (because the electrical characteristics of terminals determine the amount of calcium that enters to trigger transmitter release). This aspect of the model, then, is a descendant of earlier work on facilitation and post-tetanic potentiation at the neuromuscular junction, in which synaptic efficacy was shown to depend on altered presynaptic calcium entry.

The conclusion of most studies of learning in invertebrates, however, has been that the *ultimate* repository of long-term neural change (memory) is altered expression of neuronal genes pertinent to the efficacy of synaptic transmission (Quinn, 1984; Goelet et al., 1986). In this view, the first step in establishing a memory trace is modification of terminal membrane proteins by second messengers generated in synaptic terminals. In the short-term, second messengers change the efficacy of the synapse by covalent modification of ion channels in the terminal membrane. Such modification changes the membrane conductance and hence the amount of calcium that enters the terminal.

The long-term retention of these initial changes, however, depends on the maintained activity of the enzymes responsible for the synthesis of the relevant second messengers. This more permanent effect on second-messenger levels is postulated to occur by the activation of specific effector genes, whose purpose is to perpetuate the initial synaptic change. This strategy is in some ways akin to the activation of specific genes during cellular differentiation.

In what way, then, does evidence for the ongoing modification of connections by means of trophic interactions bear on these studies of synaptic efficacy and the idea that the ultimate locus of long-term change in the nervous system is the neuronal genome? Trophic mechanisms are unlikely to be involved in short-term modification of reflex behavior. However, the gradual revision of neural connections that trophic mechanisms enable *is* a continually updated memory of what the nervous system has learned about the body it serves, in much the same way that the connections initially laid down in development represent a memory of what a given species has learned about the circumstances it will face in the world. The rationale for this more gradual sort of learning is straightforward. The neural requirements presented by targets that vary in size and form cannot be fully known in advance, any more than extracorporeal experience can be foreseen, although the *general* demands of both the body and the world are fully anticipated in neural development. Accordingly, the purpose of learning generated by events in the external environment and learning generated by events within the body is much the same: to ensure appropriate neural function in the face of uncertain circumstances.

From this perspective, trophic interactions provide a second and quite separate cellular and molecular mechanism for learning which operates in parallel with mechanisms that alter the efficacy of existing synapses. In the case of learning based on trophic interactions there is no particular reason to think that long-term neural changes attain permanence through the activation of special genes. And there is no fundamental sense in which this sort of long-term neuronal change is encoded in the genome: the importance of genes in neural modification through trophic interactions is no more—nor less—than in any other cellular function. Permanence, such as it is in the trophic scheme of things, resides in the anatomy of patterns of connections. Such patterns represent a state of equilibrium in which the formation of new neuronal connections and the retraction of existing ones are dynamically balanced. The forces germane to this balance include the changing size and form of targets, the production of and sensitivity to trophic agents (which may vary with age, for example), the geometry of target cells, and the timing and amount of electrical activity among competing sets of terminals.

In spite of these apparent differences, there may be a link between short-term changes of synaptic efficacy induced by the action of neurotransmitters and the gradual modifications of neural connections that arise through trophic interactions. For instance, morphological studies of some of the neurons associated with modified behavioral responses in *Aplysia* indicate that active zones (distinct regions within the presynaptic terminals at which synaptic vesicles are released) change in terminals whose efficacy has been altered by activity imposed a day or two earlier (Bailey and Chen, 1983, 1985; Bailey and Kandel, 1985). The number and size of active zones is increased when the relevant synapses are shown by behavioral or electrophysiological measurement to be more effective, and vice versa. Moreover, the number of synaptic varicosities (the swellings along terminal branches which contain the active zones) also changes in parallel with long-term changes in synaptic efficacy, and it has been suggested that such changes may represent a transition of short-term changes to a more permanent form of storage (Bailey and Chen, 1983, 1986). Whether these changes are based on modulation of trophic interactions (or indeed whether trophic signals exist in invertebrates) is not yet known. This possibility, however, provides another good reason for undertaking the study of trophic interactions at this taxonomic level.

In sum, learning can occur in more than one way. One strategy involves changes in the efficacy of existing synaptic connections. Another strategy involves changes in the number and arrangement of neural connections according to trophic interactions. Rudimentary though the primary purposes of this second mechanism of learning may seem (to store and update information about the body), it is as important to the organism as learning that encodes information about external circumstances. The influence of activity on trophic effects (see Chapter 8) may ultimately provide some insight into the way these two strategies interact.

Implications for Regressive Theories of Neural Connectivity

A substantial literature has accumulated on the establishment of adult connectivity by a process of selection from a more extensive early repertoire of neural connections (Changeux et al., 1973; Changeux and Danchin, 1976; Edelman, 1978, 1981, 1982, 1985, 1987; Edelman and

Finkel, 1984; Young, 1979; Changeux et al., 1984; Ebbesson, 1984; Changeux, 1985; Toulouse et al., 1986). Evaluation of this body of work is particularly important, because much of it has been written for a wider audience of biologists and nonbiologists which, by and large, has received this idea with enthusiasm. These theories, put forward to explain the brain or mind in manifestations as diverse as perception, memory, consciousness, and free will, are necessarily different in their particulars. The theories share, however, the notion that development generates an initial set of neural connections that is ultimately reduced, anatomically or functionally, to a more restricted, permanent set by the selection of some neural circuits and the regression of others. Each of the theories also draws an explicit analogy between the development of neural connections and natural selection. This general perspective has been extended to phylogeny by the suggestion that the evolution of the nervous system proceeds by the elimination of redundant connections during the process of speciation (Ebbesson, 1984). Thus, the nervous systems of extant animals are supposed to have arisen by the selective loss of connections present in ancestral forms.

That several features of neural development have a regressive character is beyond dispute: some neurons, some connections, and some initial projections are certainly lost in the course of embryogenesis and postnatal growth. Moreover, each of these phenomena is in some sense competitive. Yet evidence for the existence of trophic interactions, their mechanism, and their purpose is difficult to reconcile with the idea that adult patterns of connections arise by the selection of some connections from an initial set and the regression of the rest.

The basic discrepancy between these ideas and the trophic theory concerns the ongoing nature of trophic change and its purposes. Theories of selection from an initial repertoire hold that a large number of connections are made initially, but that only the useful ones are retained. The burden of the trophic theory and the observations that support it is that development involves the ongoing creation of connections for a prolonged period in early life and, in many instances, on into maturity. From the trophic perspective, the repertoire of connections is never complete, and the regression of some connections is an inevitable consequence of the way in which neural adjustment occurs, connections being continually made and broken.

Why, then, has the notion of regression come to play such a prominent part in these several theories? With respect to neural connections *per se*, the apotheosis of this idea is evidently based on two observations: the apparent *net* loss of synapses in the peripheral nervous system during early development and the apparent decrease in the overall number of axonal branches and synapses in the developing brain. As for the notion of a net loss of synapses in developing muscle and ganglia, a good deal of the evidence indicates that this interpretation is simply wrong (see Chapter 5). Because several axons innervate mammalian skeletal muscle fibers which are later innervated by a single input, it would appear to follow that a surfeit of synapses is depleted by competition, in much the way that an initial pool of neurons is depleted by competition for survival at an earlier developmental stage. However, the number of terminal branches and synaptic specializations on muscle fibers actually increases during this time, just as it does in simple neuronal systems (autonomic ganglia) that can be analyzed directly. Similarly, the view that events in the developing brain imply a *net* loss of branches and synapses is open to question. Given that the axonal and dendritic processes of a fixed number of neurons continue to grow throughout the period of maturation and beyond, *rearrangement* seems a better term than *regression* to describe what happens during neural development.

Despite evidence that the idea of a net regression of neuronal branches and synapses is inappropriate as a general description of neural development, it could still be argued that development entails a net loss of *inputs,* thereby supporting the validity of a regressive principle. Certainly a net loss of inputs to target cells does occur in developing muscle, in some autonomic ganglia, and in some central pathways. However, what is known in relatively simple parts of the nervous system suggests that a *net* elimination of inputs in development is unlikely to be found among the majority of target cells in the central nervous system. Whereas a net loss of inputs from target cells is characteristic of the major type of skeletal muscle cells (fast-twitch fibers) and adendritic neurons (target cells on which the competing inputs are confined to the neuronal cell body), the addition of a few dendrites to a nerve cell diminishes the vigor of this interaction. Thus, a net reduction in the number of inputs contacting target cells is no longer observed among autonomic neurons with even moderately complex dendritic arbors (even though competitive interactions continue to determine convergence on such target cells). Similarly, muscle fibers that are incapable of generating action potentials permit multiple innervation to persist, as does synchronous activity of competing inputs. Thus, whereas *competition* among inputs as a means of regulating the number and distribution of inputs is very likely to be universal, a *net* reduction of inputs probably occurs on only the minority of target cells on which competition is made particularly intense by the confinement of asynchronously active inputs to a limited domain that is strongly depolarized by activity, such as the endplate or the neuronal cell soma.

Finally, connections in the mature nervous system are actively *maintained,* a fact revealed by experimental perturbations which elicit renewed growth or retraction of existing connections. Even in the absence of experimental intervention, ongoing remodeling of connections is apparent in those systems in which it has been possible to follow identified neural connections over time in living animals. Throughout development and on into maturity, the construction of new neuronal branches and synapses apparently happens concurrently with the removal of some pre-existing ones. Whether in the end this ongoing process reflects a net increase or net decrease in the numbers of synapses and inputs in the nervous system is less important than the rationale behind this evidence of neural malleability. It is the capacity of nerve cells continually to reorganize their connections, not the issue of net change, that is significant. To emphasize the regressive component of neural malleability (or the constructive, for that matter) misses the point. The competitive rearrangement of neural connections evident in the development of mammals and other vertebrates reflects not some abstract Darwinian principle but adjustments of neuronal branches and their connections required by elemental features of somatic development, such as the changing size and form of animals.

Since the inception of modern biology, its practitioners have sought to understand the relationship between the developmental history of individual organisms and the history of species (e.g. Gould, 1977). In a sense, theories based on the selection of neural connections from a larger initial repertoire, popularly referred to as "neural Darwinism," are the latest representatives of the attempt to interpret an aspect of ontogeny in terms of phylogeny. The facts of neural development, and the purposes of the ongoing malleability of neural connections suggested by the trophic theory, imply that the analogy between the regressive aspect of neural development and natural selection is unlikely to be any more successful than the intellectually attractive but simplistic (and ultimately erroneous) nineteenth-century proposition that ontogeny recapitulates phylogeny.

TEN

Conclusion

THE TROPHIC theory proposes that patterns of neural connections in vertebrates (and probably in most animals with nervous systems) are regulated in part by signals that derive from neural targets. Historically, this general idea stems from a series of experiments earlier in the century which established that the survival of vertebrate neurons in embryonic life depends on a competitive relationship with neuronal targets. The extension of the concept of trophic action from neuronal survival to neural connections is motivated both by an appreciation that the development and maturation of relatively large and complex animals demands much ongoing revision of connections and by the fact that such rearrangements are evident in many parts of the mammalian nervous system.

The cellular and molecular mechanisms that enable the persistent adjustment of neural connections are taken to be the same as or variants of the mechanisms that help define neuronal populations through competition for trophic support in embryonic life. At the cellular level, revision of connections occurs through the regulation of axonal and dendritic branches and terminals. At the molecular level, the trophic regulation of connections is based on the production and acquisition of specific regulatory molecules produced by target cells. Accordingly, competition can be provisionally defined as the consequence of neurons and their processes seeking an adequate share of trophic agents that are in limited supply. In those instances in which the relation between axons and targets is one of general proximity—that is, the relationship does not involve *direct* axonal contact with target cells—competition may be based on restricted amounts of diffusible trophic agents in the vicinity of the target cells. In instances in which synapses are made by direct apposition of precisely aligned pre- and postsynaptic elements, the acquisition of trophic support is probably restricted to the synaptic zone.

Among those cells that are directly related by synapses, the acquisition of trophic support involves competition between *sets* of terminals that derive from a single axon and impinge on a particular target cell. These facts suggest that competing sets are defined by the activity of the pre- and postsynaptic elements, and, more specifically, that the mechanism of this linkage may be a dependence of trophic support on simultaneous pre- and postsynaptic activation. Whether or not this speculation proves to be correct, there is no doubt that electrical activity affects trophic interactions. Decreased levels of postsynaptic activity increase the trophic stimulus provided by target cells to their innervation, and increased postsynaptic activity appears to diminish it. These effects of activity establish a regulatory loop in which the trophic responses of synaptic partners are determined interactively: the innervation of a target influences the trophic properties of the target cells, and the trophic properties of target cells influence the innervation they receive. Finally, because trophic influences are expressed throughout a chain of innervation, effects that are initiated at the level of peripheral targets can modulate the connections of an entire neural pathway.

The purposes served by trophic actions and their interactive modulation are several. The fundamental purpose of such regulation is to provide a means by which a fixed number of nerve cells can apportion innervation in a quantitatively appropriate manner during early development and thereafter adjust to the needs of a changing body throughout a prolonged period of maturation. However, the neural flexibility that arises for this purpose is likely to be used in a variety of additional ways. First, trophic interactions almost certainly play a part in the ability of the nervous system to respond to injury. Second, by generating changes in neural connections, trophic interactions provide a means whereby experience can leave a permanent trace in the nervous system. Third, trophic interactions can facilitate speciation by allowing neural adjustment to phenotypic changes favored by natural selection.

The evidence for the trophic theory is still incomplete. It would not be surprising, therefore, if some of the specific suggestions made here about the mechanisms of trophic interaction are off the mark or simply wrong. Uncertainty about particular aspects of trophic interactions, however, should not diminish the force of the general argument for trophic regulation of neural connections. There seems little alternative to the conclusion that the connections between neurons and their targets are subject to continual adjustment by means of intercellular signals. Recognition of this fact may influence the ways in which neurobiologists approach a variety of practical and theoretical issues.

Contemporary theories of perception, cognition, intelligence, and other mental attributes that are of particular interest to human beings tend to regard the fact that the brain resides in a body as a superfluous detail. Because the trophic theory is based on the responses of the nervous system to changing targets, it may serve to remind us that a first step in deciphering the brain is to understand the way it serves the needs of the body it tenants.

Glossary

action potential: the electrical signal conducted along axons (or muscle fibers) by which information is conveyed from one place to another in the nervous system

adrenergic: referring to synaptic transmission mediated by the release of norepinephrine or epinephrine

adult: the mature form of an animal, usually defined by the ability to reproduce

afferent: an axon that conducts action potentials from the periphery toward the central nervous system

anterograde: an influence acting from the neuronal cell body toward the axonal target

antiserum: serum harvested from an animal immunized to an agent of interest, such as nerve growth factor

autonomic nervous system: the part of the nervous system concerned with the activation of smooth muscle, cardiac muscle, and glands

axon: the neuronal process that carries the action potential from the nerve cell body to a target

axoplasmic transport: the process by which materials are carried from nerve cell bodies to their terminals (anterograde transport) or from nerve cell terminals to the neuronal cell body (retrograde transport)

axotomy: surgical interruption of an axon

blastomere: a cell produced when the egg undergoes cleavage

bouton: a synaptic terminal on a target cell

central nervous system: the brain and spinal cord of vertebrates; by analogy, the central nerve cord and ganglia of invertebrates

cerebellum: a division of the vertebrate brain located behind the cerebrum, concerned with motor coordination, posture, and balance

cerebrum: the largest part of the brain in man and other mammals, consisting primarily of the two cerebral hemispheres

chemoaffinity (chemoaffinity hypothesis): the idea that nerve cells bear chemical labels which determine their connectivity

cholinergic: referring to synaptic transmission mediated by the release of acetylcholine

competition: the struggle among animals, nerve cells, or nerve cell processes for limited resources essential to survival or good health

conspecific: fellow member of a species

convergence: innervation of a target cell by axons from more than one neuron

corpus callosum: a large collection of axons that unite the two cerebral hemispheres across the midline

cortex: the outer layers of the cerebral hemispheres and cerebellum, where most of the neurons in the brain are located

critical period: a limited period in the development of animals during which a behavior or some other phenomenon can be changed

dendrite: a neuronal process that receives synaptic input, usually branching near the cell body and unable to support an action potential

denervation: removal of the innervation to a target

depolarization: the displacement of a cell's membrane potential toward a less negative value, which normally triggers action potentials in excitable cells

determinate: a strategy of development in which the fates of cells and their progeny are determined at an early stage

determination: commitment of a developing cell or cell group to a particular fate

divergence: the branching of an axon to innervate multiple target cells

dorsal root: the collection of afferent nerves entering the dorsal surface of each spinal cord segment in vertebrates

dorsal root ganglia: the sensory ganglia associated with the dorsal roots

efferent: an axon that conducts information away from the central nervous system

endplate: the complex postsynaptic specialization at the site of nerve contact on skeletal muscle fibers

facilitation: the increased transmitter release produced by an action potential that follows closely upon a preceding action potential

ganglion (pl. ganglia): in vertebrates, collections of hundreds to thousands of neurons at anatomically defined sites outside the brain and spinal cord; in invertebrates, stereotyped collections of neurons that make up the "central" nervous system

ganglion cell: a neuron located in a ganglion

gastrula: the name given the early embryo during the period when its three primary layers are formed

glia (glial cells): the supportive cells associated with neurons (astrocytes and oligodendrocytes in the central nervous system, Schwann cells in peripheral nerves, and satellite cells in ganglia)

hippocampus: a deeply infolded portion of the cerebral cortex in higher vertebrates, whose function is unknown, although it has been implicated in learning and memory

higher order neurons: neurons that are relatively remote from peripheral targets

homologous: technically, referring to structures in different species that share the same evolutionary history; more generally, referring to structures or organs that have the same general anatomy and perform the same function

horseradish peroxidase: a plant enzyme widely used to stain nerve cells (after injection into a neuron, it generates a visible precipitate by one of several histochemical reactions)

induction: the ability of a cell or tissue to influence the fate of nearby cells or tissues during development, presumably by a chemical signal

innervate: establish synaptic contact with a target

innervation: referring to all the synaptic contacts made with a target

input: the innervation of a target cell by a particular axon; more loosely, the innervation of a target or target cell

input elimination: the developmental process by which the number of axons innervating some classes of target cells is diminished

interneuron: technically, a neuron in the pathway between primary sensory and primary effector neurons; in practice, a neuron that branches locally to innervate other neurons

invertebrate: an animal without a backbone (including about 97 percent of extant animals)

lateral geniculate nucleus: a nucleus in the mammalian thalamus that receives the axonal projections of retinal ganglion cells in the visual pathway from retina to cortex

long-term: lasting weeks, months, or longer

map: the ordered projection of axons from one region of the nervous system to another, by which the organization of the body is reflected in the organization of the nervous system

motor: pertaining to movement

motor cortex: the part of the cerebral cortex devoted to the organization of movement

motor neuron: a nerve cell that innervates muscle (usually restricted to skeletal muscle)

motor unit: technically, a motor neuron and the skeletal muscle fibers it innervates; more loosely, the collection of skeletal muscle fibers innervated by a single motor neuron

nematode: a roundworm or threadworm; technically, a member of the class *Nematoda*

nerve: a collection of axons, usually peripheral, that travel a common route

nerve growth factor: a small protein that acts as a trophic agent in several parts of the nervous system

neural change: a long-term change, either morphological or physiological, in the organization of the nervous system

neural connections: the axonal and dendritic branches, and the synapses between them, that link nerve cells and their targets

neural crest: cells along the dorsum of the neural tube in developing vertebrates that give rise to peripheral neurons and glia (among other derivatives)

neural unit: the collection of target neurons innervated by a single axon (analogous to motor unit)

neurite: a neuronal branch (usually used when the process in question could be either an axon or a dendrite, such as the branches of isolated nerve cells in tissue culture)

neuromuscular junction: the synapse made by a motor axon on a skeletal muscle fiber

neuron: a nerve cell, usually defined by the ability to receive and transmit information by means of synaptic connections

neuronal geometry: the spatial arrangement of neuronal branches

neuropil: the regions of neural tissue occupied by axons, dendrites, and the synapses between them

neurotransmitter: see transmitter

ocular dominance columns: complex stripes about 0.5mm wide in layer IV of the primary visual cortex of certain mammals which represent the segregation of lateral geniculate inputs from the right and left eyes

ontogeny: the developmental history of an individual animal; also used as a synonym for development

optic tectum: the first station in the visual pathway of many vertebrates, located in the midbrain and analogous to the superior colliculus in mammals

parasympathetic nervous system: a division of the peripheral autonomic nervous system in vertebrates comprising cholinergic ganglion cells located near target organs

peripheral nervous system: all nerves, neurons, and glial cells outside the central nervous system

phylogeny: the evolutionary history of a species or other taxonomic category

plasticity: long-term structural or functional changes in the nervous system

postsynaptic: referring to the component of a synapse specialized for transmitter reception

post-tetanic potentiation: the increased efficacy of synaptic transmission that follows repetitive activity of a synapse

presynaptic: referring to the component of a synapse specialized for transmitter release

primary neuron: a neuron that directly links muscles, glands, and sense organs to the central nervous system

Purkinje cell: a large neuron in the cerebellar cortex, widely used in studies of the central nervous system (named after the nineteenth-century physiologist J. E. Purkinje)

regulation: the ability of an embryonic part to compensate for partial ablation

remodeling: ongoing change in the anatomical arrangement of neural connections

resting potential: the inside-negative electrical potential that is normally found across cell membranes

retinotectal system: the pathway between ganglion cells in the retina and the optic tectum of certain vertebrates

retrograde: an influence acting from the axon terminal toward the neuronal cell body

Schwann cell: a glial cell located in the peripheral nerves (named after the nineteenth-century anatomist and physiologist T. Schwann)

secondary neuron: a neuron connected to a primary neuron and thus at one remove from direct innervation of a target

sensory: pertaining to sensation

silver stain: a classical method for visualizing neurons and their processes by impregnation with silver salts (the best-known technique is the Golgi stain, developed by the Italian anatomist C. Golgi in the late nineteenth century)

soma (pl. somata): a nerve cell body; more generally, a body

somatic cells: the cells of an animal other than its germ cells

somatosensory: denoting those parts of the nervous system involved in processing sensory information about the body itself (e.g., temperature, limb position, and mechanical deformation)

species: a taxonomic category subordinate to genus, defined in practice by anatomical similarity and the ability to interbreed

specificity: broadly, the stereotyped outcome of development in the nervous system; sometimes, selective synapse formation

spinal ganglia: dorsal root ganglia

sprouting: the growth of axon branches in response to a variety of normal or experimental stimuli

strabismus: the condition of being cross-eyed

sympathetic nervous system: a division of the peripheral autonomic nervous system in vertebrates comprising, for the most part, adrenergic ganglion cells located relatively far from the related end organs

synapse: a specialized apposition between a neuron and its target cell for transmission of information, usually involving release and reception of a chemical transmitter agent

target (neural target): the object of innervation, which can be either non-neuronal targets, such as muscles, glands, and sense organs, or other neurons

taxa: taxonomic categories

taxonomic: denoting the classification of animals

terminal: a presynaptic (axonal) ending

threshold: the level of membrane potential that generates an action potential

tonic muscle fiber: a skeletal muscle fiber which, instead of reacting to a stimulus by giving an action potential and a consequent twitch, responds to depolarization by graded contraction

transmitter (neurotransmitter): a chemical agent released by a presynaptic nerve terminal, which influences the conductance of a postsynaptic cell

trophic: the ability of one tissue or cell to support another; usually applied to long-term interactions between pre- and postsynaptic cells

trophic agent (factor): a molecule that mediates trophic interactions

trophic interactions: referring to the long-term interdependence of nerve cells and their targets

twitch muscle fiber: the major class of skeletal muscle fibers in vertebrates, which respond to a stimulus by an action potential and a rapid contraction

ventral root: the efferent nerves that exit on the ventral surface of each spinal cord segment in vertebrates

vertebrate: an animal with a backbone; technically, a member of the subphylum *Vertebrata*

visceral nervous system: the autonomic nervous system

vital dye: a reagent that stains cells only when they are alive

Bibliography

Addison, W. H. F. 1911. The development of the Purkinje cells and of the cortical layers in the cerebellum of the albino rat. *J. Comp. Neurol.* 21:459–487.

Aguayo, A. J., L. C. Terry, and G. M. Bray. 1973. Spontaneous loss of axons in sympathetic unmyelinated nerve fibers of the rat during development. *Brain Res.* 54:360–364.

Aguayo, A. J., M. Vidal-Sanz, M. P. Villegas-Pérez, and G. M. Bray. 1987. Growth and connectivity of axotomized retinal neurons in adult rats with optic nerves substituted by PNS grafts linking the eye and the midbrain. *Ann. N.Y. Acad. Sci.* 495:1–9.

Alley, K. E., and M. D. Barnes. 1983. Birth dates of trigeminal motoneurons and metamorphic reorganization of the jaw myoneural system in frogs. *J. Comp. Neurol.* 218:395–405.

Alley, K. E., and J. A. Cameron. 1983. Turnover of anuran jaw muscles during metamorphosis. *Anat. Rec.* 205:7A–8A.

Altman, J. 1963. Autoradiographic investigation of cell proliferation in the brains of rats and cats. *Anat. Rec.* 145:573–591.

——— 1972. Postnatal development of the cerebellar cortex in the rat. II. Phases in the maturation of Purkinje cells and of the molecular layer. *J. Comp. Neurol.* 145:399–464.

Altman, J., and S. A. Bayer. 1977. Time of origin and distribution of a new cell type in rat cerebellar cortex. *Exp. Brain Res.* 29:265–274.

Altman, J., and G. D. Das. 1965. Autoradiographic and histological evidence of postnatal hippocampal neurogenesis in rats. *J. Comp. Neurol.* 124:319–336.

Altman, J., and A. T. Winfree. 1977. Postnatal development of cerebellar cortex in rat. 5. Spatial organization of Purkinje cell perikarya. *J. Comp. Neurol.* 171:1–16.

Altman, P. L., and D. S. Dittmer, eds. 1961. *Blood and Other Body Fluids.* Bethesda, Md.: Federation of American Societies for Experimental Biology.

——— 1962. *Growth, Including Reproduction and Morphological Development.*

Bethesda, Md.: Federation of American Societies for Experimental Biology.

Angeletti, R. H., and R. A. Bradshaw. 1971. Nerve growth factor from mouse submaxillary gland: Amino acid sequence. *Proc. Natl. Acad. Sci.* (U.S.) 68:2417–2420.

Angeletti, R. H., M. A. Hermodson, and R. A. Bradshaw. 1973. Amino acid sequences of mouse 2.5S nerve growth factor. II. Isolation and characterization of the thermolytic and peptic peptides and the complete covalent structure. *Biochem.* 12:100–115.

Angevine, J. B. 1965. Time of neuron origin in the hippocampal region: An autoradiographic study in the mouse. *Exp. Neurol.* suppl. 2:1–70.

Angevine, J. B., and R. L. Sidman. 1962. Autoradiographic study of histogenesis in the cerebral cortex of the mouse. *Anat. Rec.* 142:210.

Anzil, A. P., A. Bieser, and A. Wernig. 1984. Light and electron microscopic identification of nerve terminal sprouting and retraction in normal adult frog muscle. *J. Physiol.* (Lond.) 350:393–399.

Arbas, E. A., and L. P. Tolbert. 1986. Presynaptic terminals persist following degeneration of "flight" muscle during development of a flightless grasshopper. *J. Neurobiology* 17:627–636.

Ariëns-Kappers, C. U., G. C. Huber, and E. C. Crosby. 1960. *The Comparative Anatomy of the Nervous System of Vertebrates, Including Man.* 3 vols. New York: Hafner.

Arnold, A. 1985. Gonadal steroid-induced organization and reorganization of neural circuits involved in bird song. In *Synaptic Plasticity,* ed. C. W. Cotman, New York: The Guilford Press, pp. 263–285.

Attardi, D. G., and R. W. Sperry. 1963. Preferential selection of central pathways by regenerating optic fibers. *Exp. Neurol.* 7:46–64.

Auburger, G., R. Heumann, R. Hellweg, S. Korsching, and H. Thoenen. 1987. Developmental changes of nerve growth factor and its mRNA in the rat hippocampus: Comparison with choline acetyltransferase. *Dev. Biol.* 120: 322–328.

Bailey, C. H., and M. Chen. 1983. Morphological basis of long-term habituation and sensitization in *Aplysia. Science* 220:91–93.

——— 1985. Morphological basis of short-term habituation in *Aplysia. Soc. Neurosci. Abstr.* 11:1110.

——— 1986. Long-term sensitization in *Aplysia* increases the total number of varicosities of single identified sensory neurons. *Soc. Neurosci. Abstr.* 12:860.

Bailey, C. H., and E. R. Kandel. 1985. Molecular approaches to the study of short and long-term memory. In *Functions of the Brain,* ed. C. W. Coen. Oxford: Oxford University Press.

Bandtlow, C. E., R. Heumann, M. E. Schwab, and H. Thoenen. 1987. Cellular localization of nerve growth factor synthesis by *in situ* hybridization. *EMBO J.* 6:891–899.

Banker, G. A., and W. M. Cowan. 1979. Further observations on hippocampal neurons in dispersed cell culture. *J. Comp. Neurol.* 187:469–494.

Barasa, A. 1960. Forma, grandezza e densità dei neuroni della corteccia cerebrale in mammiferi di grandezza corporea differente. *Z. Zellforschung* 53:69–89.

Bard, P. 1938. Studies on the cortical representation of somatic sensibility. *Bull. N.Y. Acad. Med.* 14:585–607.

Barde, Y.-A., A. M. Davies, J. E. Johnson, R. M. Lindsay, and H. Thoenen. 1987. Brain derived neurotrophic factor. In *Progress in Brain Research,* vol. 71, ed. F. J. Seil, E. Herbert, and B. M. Carlson. London: Elsevier, pp. 185–189.

Barde, Y.-A., D. Edgar, and H. Thoenen. 1982. Purification of a new neurotrophic factor from mammalian brain. *EMBO J.* 1:549–553.

——— 1983. New neurotrophic factors. *Ann. Rev. Physiol.* 45:601–612.

——— 1987. Neurotrophic factors in the central nervous system. In *Brain Peptides Update,* vol. 1, ed. J. Martin, M. Brownstein, and D. Krieger. New York: Wiley, pp. 240–249.

Barker, D., and M. C. Ip. 1966. Sprouting and degeneration of mammalian motor axons in normal and de-afferentated skeletal muscle. *Proc. R. Soc. Lond.,* ser. B, 163:538–554.

Barnes, M. D., and K. E. Alley. 1983. Maturation and recycling of trigeminal motoneurons in anuran larvae. *J. Comp. Neurol.* 218:406–414.

Baserga, R. 1985. *The Biology of Cell Reproduction.* Cambridge, Mass.: Harvard University Press.

Baulac, M., and V. Meininger. 1983. Postnatal development and cell death in the sciatic motor nucleus of the mouse. *Exp. Brain Res.* 50:107–116.

Bayer, S. A., J. W. Yackel, and P. S. Puri. 1982. Neurons in the rat dentate gyrus granular layer substantially increase during juvenile and adult life. *Science* 216:890–892.

Belford, G. R., and H. P. Killackey. 1980. The sensitive period in the development of the trigeminal system of the neonatal rat. *J. Comp. Neurol.* 193:335–350.

Bennett, M. R. 1983. Development of neuromuscular synapses. *Physiol. Rev.* 63:915–1048.

Bennett, M. R., P. A. McGrath, D. F. Davey, and I. Hutchinson. 1983. Death of motorneurons during the postnatal loss of polyneuronal innervation of rat muscles. *J. Comp. Neurol.* 218:351–363.

Bennett, M. R., and A. G. Pettigrew. 1974. The formation of synapses in striated muscle during development. *J. Physiol.* (Lond.) 241:515–545.

——— 1975. The formation of synapses in amphibian striated muscle during development. *J. Physiol.* (Lond.) 252:203–239.

Benoit, P., and J.-P. Changeux. 1975. Consequences of tenotomy on the evolution of multi-innervation in developing rat soleus muscle. *Brain Res.* 99:354–358.

——— 1978. Consequences of blocking the nerve with a local anaesthetic on the evolution of multi-innervation at the regenerating neuromuscular junction of the rat. *Brain Res.* 149:89–96.

Berg, D. K. 1984. New neuronal growth factors. *Ann. Rev. Neurosci.* 7:149–170.

Berry, M., P. McConnell, and J. Sievers. 1980. Dendritic growth and the control of neuronal form. *Current Topics in Dev. Biol.* 15:67–101.

Betz, W. J., J. H. Caldwell, and R. R. Ribchester. 1979. The size of motor units during post-natal development of rat lumbrical muscle. *J. Physiol.* (Lond.) 297:463–478.

——— 1980. The effects of partial denervation at birth on the development of muscle fibres and motor units in rat lumbrical muscle. *J. Physiol.* (Lond.) 303:265–279.

Bieser, A., A. Wernig, and H. Zucker. 1984. Different quantal responses within single frog neuromuscular junctions. *J. Physiol.* (Lond.) 350:401–412.

Birks, R., B. Katz, and R. Miledi. 1960. Physiological and structural changes at the amphibian myoneural junction, in the course of nerve degeneration. *J. Physiol.* (Lond.) 150:145–168.

Bishop, P. O. 1987. Binocular vision. In *Adler's Physiology of the Eye: Clinical Application,* ed. R. A. Moses and W. M. Hart, Jr. St. Louis: C. V. Mosby, pp. 619–683.

Bixby, J. L., and D. C. Van Essen. 1979. Regional differences in the timing of synapse elimination in skeletal muscles of the neonatal rabbit. *Brain Res.* 169:275–286.

Black, I. B. 1978. Regulation of autonomic development. *Ann. Rev. Neurosci.* 1:183–214.

Blackshaw, S. E., J. G. Nicholls, and I. Parnas. 1982. Expanded receptive fields of cutaneous mechanoreceptor cells after single neurone deletion in leech central nervous systems. *J. Physiol.* (Lond.) 326:261–268.

Blair, S. S. 1983. Blastomere ablation and the developmental origin of identified monoamine-containing neurons in the leech. *Dev. Biol.* 95:65–72.

Bliss, T. V. P., and T. Lømo. 1973. Long-lasting potentiation of synaptic transmission in the dentate area of the anaesthetized rabbit following stimulation of the perforant path. *J. Physiol.* (Lond.) 232:331–356.

Blue, M. E., and J. G. Parnavelas. 1983. The formation and maturation of synapses in the visual cortex of the rat. II. Quantitative analysis. *J. Neurocytol.* 12:697–712.

Boeke, J. 1932. Nerve endings, motor and sensory. In *Cytology and Cellular Pathology of the Nervous System,* vol. 1, ed., W. Penfield. Facsimile ed., New York: Hafner, 1965, pp. 243–315.

Bok, S. T. 1959. *Histonomy of the Cerebral Cortex.* Amsterdam: Elsevier.

Bonyhady, R. E., I. A. Hendry, C. E. Hill, and I. S. McLennan. 1980. Characterization of a cardiac muscle factor required for the survival of cultured parasympathetic neurones. *Neurosci. Letters* 18:197–201.

Bonyhady, R. E., I. A. Hendry, C. E. Hill, and D. J. Watters. 1985. An analysis of peripheral neuronal survival factors present in muscle. *J. Neurosci. Res.* 13:357–367.

Booth, C. M., and M. C. Brown. 1987. Innervated muscle fibres in partly denervated mouse muscles transiently express N-CAM. *J. Physiol.* (Lond.) 382:160P.

Booth, C. M., M. C. Brown, and S. K. Kemplay. 1987. Reduction of paralysis-induced terminal sprouting in mouse muscles with antibodies to N-CAM. *J. Physiol.* (Lond.) 390:177P.

Boothe, R. G., V. Dobson, and D. Y. Teller. 1985. Postnatal development of vision in human and non-human primates. *Ann. Rev. Neurosci.* 8:495–545.

Born, D. E., and E. W. Rubel. 1985. Afferent influences on brain stem auditory nuclei in the chicken: Neuron number and size following cochlea removal. *J. Comp. Neurol.* 231:435–445.

Bourgeois, J.-P., M. Toutant, J.-L. Gouzé, and J.-P. Changeux. 1986. The effect of activity on the selective stabilization of the motor innervation of fast muscle posterior latissimus dorsi from chick embryo. *Int. J. Dev. Neurosci.* 4:415–429.

Braford, M. R., Jr. 1986. *De gustibus non est disputandem:* A spiral center for taste in the brain of the teleost fish, *Heterotis niloticus*. *Science* 232:489–491.

Breedlove, S. M. 1986. Cellular analyses of hormone influence on motoneuronal development and function. *J. Neurobiol.* 17:157–176.

Brenner, H. R., and E. W. Johnson. 1976. Physiological and morphological effects of postganglionic axotomy on presynaptic nerve terminals. *J. Physiol.* (Lond.) 260:143–158.

Brenner, S. 1973. The genetics of behaviour. *Brit. Med. Bull.* 29:269–271.

——— 1974. The genetics of *Caenorhabditis elegans*. *Genetics* 77:71–94.

Brenowitz, E. A., and A. P. Arnold. 1986. Interspecific comparisons of the size of neural song control regions and song complexity in duetting birds: Evolutionary implications. *J. Neurosci.* 6:2875–2879.

Brodal, A. 1982. Anterograde and retrograde degeneration of nerve cells in the central nervous system. In *Histology and Histopathology of the Nervous System*, ed. W. Haymaker and R. D. Adams. Springfield, Ill.: Charles C. Thomas, pp. 276–362.

Brown, M. C. 1984. Sprouting of motor nerves in adult muscles: A recapitulation of ontogeny. *Trends in Neurosci.* 7:10–14.

Brown, M. C., R. L. Holland, and W. G. Hopkins. 1981a. Excess neuronal inputs during development. In *Development in the Nervous System*, ed. D. R. Garrod and J. D. Feldman. Cambridge: Cambridge University Press, pp. 245–262.

——— 1981b. Motor nerve sprouting. *Ann. Rev. Neurosci.* 4:17–42.

Brown, M. C., W. G. Hopkins, and R. J. Keynes. 1982. Short- and long-term effects of paralysis on the motor innervation of two different neonatal mouse muscles. *J. Physiol.* (Lond.) 329:439–450.

Brown, M. C., and R. Ironton. 1977. Motor neurone sprouting induced by prolonged tetrodotoxin block of nerve action potentials. *Nature* 265:459–461.

Brown, M. C., J. K. S. Jansen, and D. C. Van Essen. 1976. Polyneuronal innervation of skeletal muscle in new-born rats and its elimination during maturation. *J. Physiol.* (Lond.) 261:387–422.

Bueker, E. D. 1948. Implantation of tumors in the hind limb field of the embryonic chick and the developmental response of the lumbosacral nervous system. *Anat. Rec.* 102:369–390.

——— 1985. *NYU Physician*, Fall: 14–16.

Buell, S. J., and P. D. Coleman. 1980. Individual differences in dendritic growth in human aging and senile dementia. In *The Psychobiology of Aging: Problems and Perspectives*, ed. D. G. Stein. New York: Elsevier-North Holland, pp. 283–296.

Burke, R. E. 1981. Motor units: Anatomy, physiology, and functional organization. In *Handbook of Physiology. Section I: Nervous System.* Vol. 2: *Motor Control, Part 1*, ed. J. M. Brookhart and V. B. Mountcastle. Bethesda, Md.: American Physiological Society, pp. 345–422.

Calder, W. A., III. 1984. *Size, Function, and Life History.* Cambridge, Mass.: Harvard University Press.

Caldwell, J. H., and R. M. A. P. Ridge. 1983. The effects of deafferentation and spinal cord transection on synapse elimination in developing rat muscles. *J. Physiol.* (Lond.) 339:145–159.

Campbell, A. W. 1905. *Histological Studies of the Localisation of Cerebral Function.* Cambridge: Cambridge University Press.

Campenot, R. B. 1977. Local control of neurite development by nerve growth factor. *Proc. Natl. Acad. Sci.* (U.S.) 74:4516–4519.

——— 1981. Regeneration of neurites in long-term cultures of sympathetic neurons deprived of nerve growth factor. *Science* 214:579–581.

——— 1982a. Development of sympathetic neurons in compartmentalized cultures. I. Local control of neurite growth by nerve growth factor. *Dev. Biol.* 93:1–12.

——— 1982b. Development of sympathetic neurons in compartmentalized cultures. II. Local control of neurite survival by nerve growth factor. *Dev. Biol.* 93:13–21.

Cardasis, C. A., and H. A. Padykula. 1981. Ultrastructural evidence indicating reorganization at the neuromuscular junction in the normal rat soleus muscle. *Anat. Rec.* 200:41–59.

Carew, T. J., and C. L. Sahley. 1986. Invertebrate learning and memory: From behavior to molecules. *Ann. Rev. Neurosci.* 9:435–487.

Carmel, P. W., and B. M. Stein. 1969. Cell changes in sensory ganglia following proximal and distal nerve section in the monkey. *J. Comp. Neurol.* 135:145–166.

Cashman, N. R., R. Maselli, R. L. Wollmann, R. Roos, R. Simon, and J. P. Antel. 1987. Late denervation in patients with antecedent paralytic poliomyelitis. *N. Engl. J. Med.* 317:7–12.

Castellucci, V. F., T. J. Carew, and E. R. Kandel. 1978. Cellular analysis of long-term habituation of the gill-withdrawal reflex of *Aplysia californica. Science* 202:1306–1308.

Castellucci, V. F., and E. R. Kandel. 1974. A quantal analysis of the synaptic depression underlying habituation of the gill-withdrawal reflex in *Aplysia. Proc. Natl. Acad. Sci.* (U.S.) 71:5004–5008.

Castellucci, V. F., H. Pinsker, I. Kupfermann, and E. R. Kandel. 1970. Neuronal mechanisms of habituation and dishabituation of the gill-withdrawal reflex in *Aplysia. Science* 167:1745–1748.

Cerf, J. A., and L. W. Chacko. 1958. Retrograde reaction in motoneuron dendrites following ventral root section in the frog. *J. Comp. Neurol.* 109:205–219.

Chalfie, M. 1984. Neuronal development in *Caenorhabditis elegans. Trends in Neurosci.* 9:197–202.

Chalupa, L. M., R. W. Williams, and Z. Henderson. 1984. Binocular interaction in the fetal cat regulates the size of the ganglion cell population. *Neuroscience* 12:1139–1146.

Changeux, J.-P. 1985. *Neuronal Man: The Biology of Mind.* Transl. by L. Garey. New York: Pantheon.

Changeux, J.-P., P. Courrège, and A. Danchin. 1973. A theory of the epigenesis of neuronal networks by selective stabilization of synapses. *Proc. Natl. Acad. Sci.* (U.S.) 70:2974–2978.

Changeux, J.-P., and A. Danchin. 1976. Selective stabilisation of developing synapses as a mechanism for the specification of neuronal networks. *Nature* 264:705–712.

Changeux, J.-P., T. Heidmann, and P. Patte. 1984. Learning by selection. In *The Biology of Learning,* ed. P. Marler and H. S. Terrace. Berlin: Springer-Verlag, pp. 115–133.

Chun, L. L. Y., and P. H. Patterson. 1977. Role of nerve growth factor in the development of rat sympathetic neurons in vitro. I. Survival, growth, and differentiation of catecholamine production. *J. Cell Biol.* 75:694–704.

Chung, K., and R. E. Coggeshall. 1987. Postnatal development of the rat dorsal funiculus. *J. Neurosci.* 7:972–977.

Chung, S. H., M. J. Keating, and T. V. P. Bliss. 1974. Functional synaptic relations during the development of the retino-tectal projection in amphibians. *Proc. R. Soc. Lond.*, ser. B, 187:449–459.

Chu-Wang, I.-W., and R. W. Oppenheim. 1978. Cell death of motoneurons in the chick embryo spinal cord. II. A quantitative and qualitative analysis of degeneration in the ventral root, including evidence for axon outgrowth and limb innervation prior to cell death. *J. Comp. Neurol.* 177:59–85.

Cobb, S. 1965. Brain size. *Arch. Neurol.* 12:555–561.

Cohen, S. 1959. Purification and metabolic effects of a nerve growth-promoting protein from snake venom. *J. Biol. Chem.* 234:1129–1137.

——— 1960. Purification of a nerve-growth promoting protein from the mouse salivary gland and its neuro-cytotoxic antiserum. *Proc. Natl. Acad. Sci.* (U.S.) 46:302–311.

Cohen, S., and R. Levi-Montalcini. 1956. A nerve growth-stimulating factor isolated from snake venom. *Proc. Natl. Acad. Sci.* (U.S.) 42:571–574.

Cohen, S., R. Levi-Montalcini, and V. Hamburger. 1954. A nerve growth-stimulating factor isolated from sarcomas 37 and 180. *Proc. Natl. Acad. Sci.* (U.S.) 40:1014–1018.

Collins, F., and K. A. Crutcher. 1985. Neurotrophic activity in the adult rat hippocampal formation: Regional distribution and increase after septal lesion. *J. Neurosci.* 5:2809–2814.

Conklin, E. G. 1911. Body size and cell size. *J. Morphol.* 23:159–188.

Cook, W. H., J. H. Walker, and M. L. Barr. 1951. A cytological study of transneuronal atrophy in the cat and rabbit. *J. Comp. Neurol.* 94:267–291.

Coppoletta, J. M., and S. B. Wolbach. 1933. Body length and organ weight of infants and children. *Am. J. Pathol.* 9:55–70.

Cotman, C. W., M. Nieto-Sampedro, and E. W. Harris. 1981. Synapse replacement in the nervous system of adult vertebrates. *Physiol. Rev.* 61:684–784.

Count, E. W. 1947. Brain and body weight in man: Their antecedents in growth and evolution. *Ann. N.Y. Acad. Sci.* 46:993–1122.

Courtney, K., and S. Roper. 1976. Sprouting of synapses after partial denervation of frog cardiac ganglion. *Nature* 259:317–319.

Covault, J., and J. R. Sanes. 1986. Distribution of N-CAM in synaptic and extrasynaptic portions of developing and adult skeletal muscle. *J. Cell Biol.* 102:716–730.

Cowan, W. M. 1970. Anterograde and retrograde transneuronal degeneration in the central and peripheral nervous system. In *Contemporary Research Methods in Neuroanatomy,* ed. S. Ebbesson and W. J. H. Nauta. New York: Springer-Verlag, pp. 217–251.

Cowan, W. M., and P. G. H. Clarke. 1976. The development of the isthmo-optic nucleus. *Brain Behav. Evol.* 13:345–375.

Cragg, B. G. 1975. The development of synapses in the visual system of the cat. *J. Comp. Neurol.* 160:147–166.

Crepel, F., N. Delhaye-Bouchaud, J. M. Guastavino, and I. Sampaio. 1980. Multiple innervation of cerebellar Purkinje cells by climbing fibres in *staggerer* mutant mouse. *Nature* 283:483–484.

Crepel, F., J. Mariani, and N. Delhaye-Bouchaud. 1976. Evidence for a multiple innervation of Purkinje cells by climbing fibers in the immature rat cerebellum. *J. Neurobiol.* 7:567–578.

Crespo, D., D. D. M. O'Leary, and W. M. Cowan. 1985. Changes in the numbers of optic nerve fibers during late prenatal and postnatal development in the albino rat. *Dev. Brain Res.* 19:129–134.

Crick, F. 1982. Do dendritic spines twitch? *Trends in Neurosci.* 5:44–46.

Cunningham, T. J. 1982. Naturally occurring neuron death and its regulation by developing neural pathways. *Int. Rev. Cytol.* 74:163–186.

Dagg, A. I., and J. B. Foster. 1976. *The Giraffe: Its Biology, Behavior, and Ecology.* New York: Van Nostrand Reinhold.

Davies, A. M. 1986. The survival and growth of embryonic proprioceptive neurons is promoted by a factor present in skeletal muscle. *Dev. Biol.* 115:56–67.

Davies, A. M., C. Bandtlow, R. Heumann, S. Korsching, H. Rohrer, and H. Thoenen. 1987. Timing and site of nerve growth factor synthesis in developing skin in relation to innervation and expression of the receptor. *Nature* 326:353–358.

Davies, A. M., H. Thoenen, and Y.-A. Barde. 1986a. Different factors from the central nervous system and periphery regulate the survival of sensory neurones. *Nature* 319:497–499.

——— 1986b. The response of chick sensory neurons to brain-derived neurotrophic factor. *J. Neurosci.* 6:1897–1904.

DeGroot, D., and G. Vrensen. 1978. Postnatal development of synaptic contact zones in the visual cortex of rabbits. *Brain Res.* 147:362–369.

Deitch, J. S., and E. W. Rubel. 1984. Afferent influences on brain stem auditory nuclei of the chicken: Time course and specificity of dendritic atrophy following deafferentation. *J. Comp. Neurol.* 229:66–79.

DeJongh, H. 1968. Functional morphology of the jaw apparatus of larval and metamorphosing *Rana temporaria. Neth. J. Zool.* 18:1–103.

Dekaban, A. S., and D. Sadowsky. 1978. Changes in brain weights during the span of human life: Relation of brain weights to body heights and body weights. *Annals of Neurol.* 4:345–356.

Del Castillo, J., and B. Katz. 1954. Statistical factors involved in neuromuscular facilitation and depression. *J. Physiol.* (Lond.) 124:574–585.

Dennis, M. J., and P. B. Sargent. 1979. Loss of extrasynaptic acetylcholine sensitivity upon reinnervation of parasympathetic ganglion cells. *J. Physiol.* (Lond.) 289:263–275.

Dennis, M. J., L. Ziskind-Conhaim, and A. J. Harris. 1981. Development of neuromuscular junctions in rat embryos. *Dev. Biol.* 81:266–279.

Detwiler, S. R. 1920. On the hyperplasia of nerve centers resulting from excessive peripheral loading. *Proc. Natl. Acad. Sci.* (U.S.) 6:96–101.

——— 1936. *Neuroembryology: An Experimental Study.* New York: Macmillan.

DeVoogd, T. J., and F. Nottebohm. 1981a. Gonadal hormones induce dendritic growth in the adult avian brain. *Science* 214:202–204.

——— 1981b. Sex differences in dendritic morphology of a song control nucleus in the canary: A quantitative Golgi study. *J. Comp. Neurol.* 196:309–316.

Diamond, J., E. Cooper, C. Turner, and L. MacIntyre. 1976. Trophic regulation of nerve sprouting. *Science* 193:371–377.

Diamond, J., M. Coughlin, L. MacIntyre, M. Holmes, and B. Visheau. 1987. Evidence that endogenous β nerve growth factor is responsible for the collateral sprouting, but not the regeneration, of nociceptive axons in adult rats. *Proc. Natl. Acad. Sci.* (U.S.) 84:6596–6600.

Dichter, M. A. 1978. Rat cortical neurons in cell culture: Culture methods, cell morphology, electrophysiology, and synapse formation. *Brain Res.* 149:279–293.

Doe, C. Q., J. Y. Kuwada, and C. S. Goodman. 1985. From epithelium to neuroblasts to neurons: The role of cell interactions and cell lineage during insect neurogenesis. *Phil. Trans. R. Soc. Lond.*, ser. B, 312:67–81.

Dohrmann, U., D. Edgar, M. Sendtner, and H. Thoenen. 1986. Muscle-derived factors that support survival and promote fiber outgrowth from embryonic chick spinal motor neurons in culture. *Dev. Biol.* 118:209–221.

Donaldson, H. H. 1895. *Growth of the Brain: Study of the Nervous System in Relation to Education.* New York: Scribner's Sons.

Downman, C. B. B., J. C. Eccles, and A. K. McIntyre. 1953. Functional changes in chromatolysed motoneurones. *J. Comp. Neurol.* 98:9–36.

Dribin, L. B., and J. N. Barrett. 1980. Conditioned medium enhances neuritic outgrowth from rat spinal cord explants. *Dev. Biol.* 74:184–195.

——— 1982. Two components of conditioned medium increase neuritic outgrowth from rat spinal cord explants. *J. Neurosci. Res.* 8:271–280.

Duffy, C. J., and P. Rakic. 1983. Differentiation of granule cell dendrites in the dentate gyrus of the rhesus monkey: A quantitative Golgi study. *J. Comp. Neurol.* 214:224–237.

Dunn, P. M., and L. M. Marshall. 1985. Lack of nicotinic supersensitivity in frog sympathetic neurones following denervation. *J. Physiol.* (Lond.) 363:211–225.

Easter, S. S., Jr., D. Purves, P. Rakic, and N. C. Spitzer. 1985. The changing view of neural specificity. *Science* 230:507–511.

Easter, S. S., Jr., and C. A. O. Stuermer. 1984. An evaluation of the hypothesis of shifting terminals in goldfish optic tectum. *J. Neurosci.* 4:1052–1063.

Ebbesson, S. O. E. 1963. A quantitative study of human superior cervical sympathetic ganglia. *Anat. Rec.* 146:353–356.

——— 1968a. Quantitative studies of superior cervical sympathetic ganglia in a variety of primates including man. I. The ratio of preganglionic fibers to ganglionic neurons. *J. Morphol.* 124:117–131.

——— 1968b. Quantitative studies of superior cervical sympathetic ganglia in a variety of primates including man. II. Neuronal packing density. *J. Morphol.* 124:181–185.

——— 1984. Evolution and ontogeny of neural circuits. *Behav. Brain Sci.* 7:321–366.

Ebendal, T., M. Belew, C.-O. Jacobson, and J. Porath. 1979. Neurite outgrowth elicited by embryonic chick heart: Partial purification of the active factor. *Neurosci. Letters* 14:91–95.

Ebendal, T., L. Olson, A. Seiger, and K.-O. Hedlund. 1980. Nerve growth factors in the rat iris. *Nature* 286:25–28.

Eccles, J. C. 1960. The properties of the dendrites. In *Structure and Function of the Cerebral Cortex,* ed. D. B. Tower and J. P. Schadé. New York: Elsevier, pp. 192–202.

——— 1986. Chromatolysis of neurones after axon section. In *Recent Achievements in Restorative Neurology 2: Progressive Neuromuscular Diseases,* ed. M. R. Dimitrijevic, B. A. Kakulas, and G. Vrbová. Basel: Karger, pp. 318–331.

Eccles, J. C., B. Libet, and R. R. Young. 1958. The behaviour of chromatolysed motoneurones studied by intracellular recording. *J. Physiol.* (Lond.) 143:11–40.

Eccles, J. C., R. Llinás, and K. Sasaki. 1966. The excitatory synaptic action of climbing fibres on the Purkinje cells of the cerebellum. *J. Physiol.* (Lond.) 182:268–296.

Edds, M. V. 1953. Collateral nerve regeneration. *Quart. Rev. Biol.* 28:260–276.

Edelman, G. M. 1978. Group selection and phasic reentrant signalling: A theory of higher brain function. In *The Mindful Brain: Cortical Organization and the Group-Selective Theory of Higher Brain Function,* ed. G. M. Edelman and V. B. Mountcastle. Cambridge, Mass.: MIT Press, pp. 55–100.

——— 1981. Group selection as the basis for higher brain function. In *Organization of the Cerebral Cortex,* ed. F. O. Schmitt, F. G. Worden, G. Adelman, and S. G. Dennis. Cambridge, Mass.: MIT Press, pp. 51–100.

——— 1982. Through a computer darkly: Group selection and higher brain function. *Bull. Am. Acad. Arts and Sciences* 36:20–49.

——— 1983. Cell adhesion molecules. *Science* 219:450–457.

——— 1985. Neural Darwinism: Population thinking and higher brain function. In *How We Know*, ed. M. Shafto. San Francisco: Harper & Row, pp. 1–30.

——— 1986. Cell adhesion molecules in the regulation of animal form and tissue pattern. *Ann. Rev. Cell Biol.* 2:81–116.

——— 1987. *Neural Darwinism: The Theory of Neuronal Group Selection.* New York: Basic Books.

Edelman, G. M., and L. H. Finkel. 1984. Neuronal group selection in the cerebral cortex. In *Dynamic Aspects of Neocortical Function*, ed. G. M. Edelman, W. M. Cowan, and W. E. Gall. New York: Wiley, pp. 653–695.

Ellis, H. M., and H. R. Horvitz. 1986. Genetic control of programmed cell death in the nematode *C. elegans*. *Cell* 44:817–829.

Faber, D. S. 1984. Reorganization of neuronal membrane properties following axotomy. *Exp. Brain Res.* suppl. 9:225–239.

Finn, P. J., I. A. Ferguson, F. J. Renton, and R. A. Rush. 1986. Nerve growth factor immunohistochemistry and biological activity in the rat iris. *J. Neurocytol.* 15:169–176.

Fischbach, G. D., E. Frank, T. M. Jessell, L. L. Rubin, and S. M. Schuetze. 1979. Accumulation of acetylcholine receptors and acetylcholinesterase at newly formed nerve-muscle synapses. *Pharmacol. Rev.* 30:411–428.

Fishman, R. B., and S. M. Breedlove. 1987. Androgen blockade of bulbocavernosus muscle inhibits testosterone-dependent masculinization of spinal motoneurons in newborn female rats. *Soc. Neur. Abstr.* 13:1520.

Fixsen, W., P. Sternberg, H. Ellis, and R. Horvitz. 1985. Genes that affect cell fates during the development of *Caenorhabditis elegans*. *Cold Spring Harbor Symp. Quant. Biol.* 50:99–104.

Fladby, T. 1987. Postnatal loss of synaptic terminals in the normal mouse soleus muscle. *Acta Physiol. Scand.* 129:229–238.

Fladby, T., and J. K. S. Jansen. 1987. Postnatal loss of synaptic terminals in partially denervated mouse soleus muscles. *Acta Physiol. Scand.* 129:239–246.

Forehand, C. J. 1985. Density of somatic innervation on mammalian autonomic ganglion cells is inversely related to dendritic complexity and preganglionic convergence. *J. Neurosci.* 5:3403–3408.

——— 1987. Ultrastructural analysis of the distribution of synaptic boutons from labeled preganglionic axons on rabbit ciliary neurons. *J. Neurosci.* 7:3274–3281.

Forehand, C. J., and D. Purves. 1984. Regional innervation of rabbit ciliary ganglion cells by the terminals of preganglionic axons. *J. Neurosci.* 4:1–12.

Forger, N. G., and S. M. Breedlove. 1986. Sexual dimorphism in human and canine spinal cord: Role of early androgen. *Proc. Nat. Acad. Sci.* (U.S.) 83:7527–7531.

Fraser, S. E. 1985. Cell interactions involved in neuronal patterning: An experimental and theoretical approach. In *Molecular Bases of Neural Development*, ed. G. M. Edelman, W. E. Gall, and W. M. Cowan. New York: Wiley, pp. 481–507.

Fraser, S. E., and M.-M. Poo. 1982. Development, maintenance, and modulation of patterned membrane topography: Models based on the acetylcholine receptor. *Current Topics in Dev. Biol.* 17:77–100.

Fridén, J. 1983. Exercise-induced muscle soreness: A qualitative and quantitative study of human muscle morphology and function. *Umeå Univ. Med. Disserts.*, n.s. 105.

——— 1984. Changes in human skeletal muscle induced by long-term exercise. *Cell Tissue Res.* 236:365–372.

Fridén, J., M. Sjöström, and B. Ekblom. 1983. Myofibrillar damage following intense eccentric exercise in man. *Int. J. Sports Med.* 4:170–176.

Fritsch, G., and E. Hitzig. 1870. The electrical excitability of the cerebrum. Trans. K. H. Wilkins, *J. Neurosurg.* 20 (1963):904–916.

Fujisawa, H. 1981. Retinotopic analysis of fiber pathways in the regenerating retinotectal system of the adult newt *Cynops pyrrhogaster. Brain Res.* 206: 27–37.

Fujisawa, H., K. Watanabe, N. Tani, and Y. Ibata. 1981. Retinotopic analysis of fiber pathways in amphibians. II. The frog *Rana nigromaculata. Brain Res.* 206:21–26.

Furber, S., R. W. Oppenheim, and D. Prevette. 1987. Naturally-occurring neuron death in the ciliary ganglion of the chick embryo following removal of preganglionic input: Evidence for the role of afferents in ganglion cell survival. *J. Neurosci.* 7:1816–1832.

Gabella, G. 1976. *Structure of the Autonomic Nervous System.* London: Chapman and Hall.

Gage, F. H., K. Wictorin, W. Fischer, L. R. Williams, S. Varon, and A. Björklund. 1986. Life and death of cholinergic neurons in the septal and diagonal band region following complete fimbria fornix transection. *Neurosci.* 19:241–256.

Gallego, R., and E. Geijo. 1987. Chronic block of the cervical trunk increases synaptic efficacy in the superior and stellate ganglia of the guinea-pig. *J. Physiol.* (Lond.) 382:449–462.

Garey, L. J., and C. de Courten. 1983. Structural development of the lateral geniculate nucleus and visual cortex in monkey and man. *Behav. Brain Res.* 10:3–13.

Gaze, R. M., M. Jacobson, and G. Székely. 1963. The retino-tectal projection in *Xenopus* with compound eyes. *J. Physiol.* (Lond.) 165:484–499.

——— 1965. On the formation of connexions by compound eyes in *Xenopus. J. Physiol.* (Lond.) 176:409–417.

Gaze, R. M., M. J. Keating, and S.-H. Chung. 1974. The evolution of the retinotectal map during development in *Xenopus. Proc. R. Soc. Lond.*, ser. B, 185:301–330.

Gaze, R. M., M. J. Keating, A. Ostberg, and S.-H. Chung. 1979. The relationship between retinal and tectal growth in larval *Xenopus:* Implications for the development of the retino-tectal projection. *J. Embryol. Exp. Morphol.* 53:103–143.

Gaze, R. M., and S. C. Sharma. 1970. Axial differences in the reinnervation of the goldfish optic tectum by regenerating optic nerve fibres. *Exp. Brain Res.* 10:171–181.

Geschwind, N. 1970. The organization of language and the brain. *Science* 170:940–944.

Glover, J. C., and A. Mason. 1986. Morphogenesis of an identified leech neuron: Segmental specification of axonal outgrowth. *Dev. Biol.* 115:256–260.

Goelet, P., V. F. Castellucci, S. Schachter, and E. R. Kandel. 1986. The long and the short of long-term memory—a molecular framework. *Nature* 322:419–422.

Goldman-Rakic, P. S., and P. Rakic. 1984. Experimental modification of gyral patterns. In *Cerebral Dominance: The Biological Foundations,* ed. N. Geschwind and A. M. Galaburda. Cambridge, Mass.: Harvard University Press.

Gordon, T., M. E. M. Kelly, E. J. Sanders, J. Shapiro, and P. A. Smith. 1987. The effects of axotomy on bullfrog sympathetic neurones. *J. Physiol.* (Lond.) 392:213–229.

Gorin, P. D., and E. M. Johnson. 1979. Experimental autoimmune model of nerve growth factor deprivation: Effects on developing peripheral sympathetic and sensory neurons. *Proc. Natl. Acad. Sci.* (U.S.) 76:5382–5386.

Gorio, A., G. Carmignoto, M. Finesso, P. Polato, and M. G. Nunzi. 1983. Muscle reinnervation. II. Sprouting, synapse formation, and repression. *Neuroscience* 8:403–416.

Goshgarian, H. G., J. M. Koistinen, and E. R. Schmidt. 1983. Cell death and changes in the retrograde transport of horseradish peroxidase in rubrospinal neurons following spinal cord hemisection in the adult rat. *J. Comp. Neurol.* 214:251–257.

Gould, S. J. 1975. Allometry in primates, with emphasis on scaling and the evolution of the brain. *Contrib. Primatol.* 5:244–292.

——— 1977. *Ontogeny and Phylogeny.* Cambridge, Mass.: Harvard University Press.

——— 1981. *The Mismeasure of Man.* New York: Norton.

Govind, C. K. 1984. Developmental asymmetry in the neuromuscular system of lobster claws. *Biol. Bull.* 167:94–119.

Govind, C. K., and R. A. Derosa. 1983. Fine structure of comparable synapses in mature and larval lobster muscle. *Tissue and Cell* 15:97–106.

Govind, C. K., and J. Pearce. 1981. Remodeling of multiterminal innervation by nerve terminal sprouting in an identifiable lobster motoneuron. *Science* 212:1522–1524.

Graf, W., and R. Baker. 1985a. The vestibuloocular reflex of the adult flatfish. I. Oculomotor organization. *J. Neurophysiol.* 54:887–899.

——— 1985b. The vestibuloocular reflex of the adult flatfish. II. Vestibulooculomotor connectivity. *J. Neurophysiol.* 54:900–916.

Grant, G. 1965. Degenerative changes in dendrites following axonal transection. *Experientia* 21:722.

——— 1968. Silver impregnation of degenerating dendrites, cells, and axons central to axonal transection. II. A Nauta study on spinal motor neurones in kittens. *Exp. Brain Res.* 6:284.

Grant, G., and J. Westman. 1969. The lateral cervical nucleus in the cat. IV. A light and electron microscopical study after midbrain lesions with demonstration of indirect Wallerian degeneration at the ultrastructural level. *Exp. Brain Res.* 7:51.

Grant, S., and M. J. Keating. 1986. Normal maturation involves systematic changes in binocular visual connections in *Xenopus laevis*. *Nature* 322:258–261.

Graziadei, P. P. C., and G. A. Monti Graziadei. 1978. Continuous nerve cell renewal in the olfactory system. In *Handbook of Sensory Physiology*, vol. 9, ed. M. Jacobson. Berlin: Springer-Verlag, pp. 55–83.

Greene, L., and E. Shooter. 1980. Nerve growth factor: Biochemistry, synthesis, and mechanism of action. *Ann. Rev. Neurosci.* 3:353–402.

Griffin, D. R. 1974. *Listening in the Dark: The Acoustic Orientation of Bats and Men.* New York: Dover Publications.

Grinnell, A. D. 1963. The neurophysiology of audition in bats. *J. Physiol.* (Lond.) 167:38–127.

Grinnell, A. D., and A. A. Herrera. 1981. Specificity and plasticity of neuromuscular connections: Long-term regulation of motoneuron function. *Prog. Neurobiol.* 17:203–282.

Guillery, R. W. 1973a. The effect of lid suture upon the growth of cells in the dorsal lateral geniculate nucleus of kittens. *J. Comp. Neurol.* 148:412–422.

——— 1973b. Quantitative studies of transneuronal atrophy in the dorsal lateral geniculate nucleus of cats and kittens. *J. Comp. Neurol.* 149:423–438.

Gunderson, R. W., and J. N. Barrett. 1980. Characterization of the turning response of dorsal root neurites toward nerve growth factor. *J. Cell Biol.* 87:546–554.

Gurney, M. E. 1981. Hormonal control of cell form and number in the zebra finch song system. *J. Neurosci.* 1:658–673.

——— 1984. Suppression of sprouting at the neuromuscular junction by immune sera. *Nature* 307:546–548.

Gurney, M. E., and M. Konishi. 1980. Hormone-induced sexual differentiation of brain and behavior in zebra finches. *Science* 208:1380–1383.

Guth, L., and J. J. Bernstein. 1961. Selectivity in the re-establishment of synapses in the superior cervical sympathetic ganglion of the cat. *Exp. Neurol.* 4:59–69.

Gutmann, E., and J. Z. Young. 1944. Reinnervation of muscle after various periods of atrophy. *J. Anat.* 78:15–43.

Haldane, J. B. S. 1985. *On Being the Right Size and Other Essays.* Oxford: Oxford University Press.

Hamburger, V. 1934. The effects of wing bud extirpation on the development of the central nervous system of chick embryos. *J. Exp. Zool.* 68:449–494.

——— 1939a. The development and innervation of transplanted limb primordia of chick embryos. *J. Exp. Zool.* 80:347–385.

——— 1939b. Motor and sensory hyperplasia following limb-bud transplantations in chick embryos. *Physiol. Zool.* 12:268–284.

——— 1958. Regression versus peripheral control of differentiation in motor hypoplasia. *Am. J. Anat.* 102:365–409.

——— 1975. Cell death in the development of the lateral motor column of the chick embryo. *J. Comp. Neurol.* 160:535–546.

——— 1977. The developmental history of the motor neuron. The F. O. Schmitt Lecture in Neuroscience, 1976. *Neurosci. Res. Program Bull.* 15 (suppl. 3):1–37.

——— 1988. *The Heritage of Experimental Embryology: Hans Spemann and the Organizer.* New York: Oxford University Press.

Hamburger, V., and R. Levi-Montalcini. 1949. Proliferation, differentiation, and degeneration in the spinal ganglia of the chick embryo under normal and experimental conditions. *J. Exp. Zool.* 111:457–501.

——— 1950. Some aspects of neuroembryology. In *Genetic Neurology*, ed. P. A. Weiss. Chicago: University of Chicago Press, pp. 128–160.

Hamburger, V., and R. W. Oppenheim. 1982. Naturally occurring neuronal death in vertebrates. *Neurosci. Comment.* 1:39–55.

Hamburger, V., and J. W. Yip. 1984. Reduction of experimentally induced neuronal death in spinal ganglia of the chick embryo by nerve growth factor. *J. Neurosci.* 4:767–774.

Hanna, C., and J. E. O'Brien. 1960. Cell production and migration in the epithelial layer of the cornea. *Arch. Ophth.* 64:536–539.

Hardesty, I. 1902. Observations on the medulla spinalis of the elephant with some comparative studies of the *intumescentia cervicalis* and the neurones of the *columna anterior*. *J. Comp. Neurol.* 12:125–182.

Hardman, V. J., and M. C. Brown. 1985. Absence of postnatal death among motoneurones supplying the inferior gluteal nerve of the rat. *Dev. Brain Res.* 19:1–19.

Harris, A. J. 1974. Inductive functions of the nervous system. *Ann. Rev. Physiol.* 36:251–305.

——— 1981. Embryonic growth and innervation of rat skeletal muscles. *Phil. Trans. R. Soc. Lond.*, ser. B, 293:257–314.

Harris, L. 1987. Dynamic aspects of corneal innervation visualized in living mice. *Soc. Neurosci. Abstr.* 13:1007.

Harvey, P. H., and P. M. Bennett. 1983. Brain size, energetics, ecology and life history patterns. *Nature* 306:314–315.

Harwerth, R. S., E. L. Smith, G. C. Duncan, M. L. J. Crawford, and G. K. Von Noorden. 1986. Multiple sensitive periods in the development of the primate visual system. *Science* 232:235–238.

Heathcote, R. D., and P. B. Sargent. 1985. Loss of supernumerary axons during neuronal morphogenesis. *J. Neurosci.* 5:1940–1946.

Hebb, D. O. 1949. *The Organization of Behavior.* New York: Wiley.

Hefti, F. 1986. Nerve growth factor promotes survival of septal cholinergic neurons after fimbrial transections. *J. Neurosci.* 6:2155–2162.

Henderson, C. E., M. Huchet, and J.-P. Changeux. 1981. Neurite outgrowth from embryonic chicken spinal neurons is promoted by media conditioned by muscle cells. *Proc. Natl. Acad. Sci.* (U.S.) 78:2625–2629.

Hendry, I. A. 1975. The response of adrenergic neurons to axotomy and nerve growth factor. *Brain Res.* 94:87–97.

——— 1977. Cell division in the developing sympathetic nervous system. *J. Neurocytol.* 6:299–309.

Hendry, I. A., and J. Campbell. 1976. Morphometric analysis of rat superior cervical ganglion after axotomy and nerve growth factor treatment. *J. Neurocytol.* 5:351–360.

Hendry, S. H. C., and E. G. Jones. 1986. Reduction in number of immunostained GABAergic neurones in deprived-eye dominance columns in monkey area 17. *Nature* 320:750–753.

Henneman, E., and L. M. Mendell. 1981. Functional organization of motoneuron pool and its inputs. In *Handbook of Physiology,* sect. 1, vol. 2: *The Nervous System,* ed. S. R. Greiger. Bethesda, Md.: American Physiological Society, pp. 423–507.

Henneman, E., G. Somjen, and D. O. Carpenter. 1965. Functional significance of cell size in spinal motoneurons. *J. Neurophysiol.* 28:560–580.

Herrera, A. A., and L. R. Banner. 1987. Direct observation of motor nerve terminal remodelling in living frogs. *Soc. Neurosci. Abstr.* 13:1665.

Herrera, A. A., and D. R. Scott. 1985. Motor axon sprouting in frog sartorius muscles is not altered by contralateral axotomy. *J. Neurocytol.* 14:145–156.

Herrup, K., A. Letsou, and T. J. Diglio. 1984. Cell lineage relationships in the development of the mammalian CNS: The facial nerve nucleus. *Dev. Biol.* 103:329–336.

Herrup, K., and K. Sunter. 1987. Numerical matching during cerebellar development: Quantitative analysis of granule cell death in *staggerer* mouse chimeras. *J. Neurosci.* 7:829–836.

Heumann, R., S. Korsching, C. Bandtlow, and H. Thoenen. 1987. Changes of nerve growth factor synthesis in nonneuronal cells in response to sciatic nerve transection. *J. Cell Biol.* 104:1623–1631.

Heumann, R., S. Korsching, J. Scott, and H. Thoenen. 1984. Relationship between levels of nerve growth factor and its messenger RNA in sympathetic ganglia and peripheral target tissues. *EMBO J.* 3:3183–3190.

Heumann, R., and H. Thoenen. 1986. Comparison between the time course of changes in nerve growth factor protein levels and those of its messenger RNA in the cultured rat iris. *J. Biol. Chem.* 261:9246–9249.

Hikida, R. S., R. S. Staron, F. C. Hagerman, W. M. Sherman, and D. L. Costill. 1983. Muscle fiber necrosis associated with human marathon runners. *J. Neurol. Sci.* 59:185–203.

Hinds, J. W., and N. A. McNelly. 1977. Aging of the rat olfactory bulb: Growth and atrophy of constituent layers and changes in size and number of mitral cells. *J. Comp. Neurol.* 171:345–368.

Hofer, M. M., and Y.-A. Barde. 1988. Brain-derived neurotrophic factor prevents neuronal death *in vivo. Nature* 331:261–262.

Holland, R. L., and M. C. Brown. 1980. Postsynaptic transmission block can cause terminal sprouting of a motor nerve. *Science* 207:649–651.

Hollyday, M., and V. Hamburger. 1976. Reduction of the naturally occurring motor neuron loss by enlargement of the periphery. *J. Comp. Neurol.* 170:311–320.

Hollyday, M., V. Hamburger, and J. M. G. Farris. 1977. Localization of motor neuron pools supplying identified muscles in normal and supernumerary legs of chick embryo. *Proc. Natl. Acad. Sci.* (U.S.) 74:3582–3586.

Hopkins, W. G., M. C. Brown, and R. J. Keynes. 1985. Post-natal growth of motor nerve terminals in muscles of the mouse. *J. Neurocytol.* 14:525–540.

Horch, K. 1981. Absence of functional collateral sprouting of mechanoreceptor axons into denervated areas of mammalian skin. *Exp. Neurol.* 74:313–318.

Horder, T. J. 1971. Retention by fish optic nerve fibres regenerating to new terminal sites in the tectum of "chemospecific" affinity for their original sites. *J. Physiol.* (Lond.) 216:53–55.

Horvitz, H. R. 1981. Neuronal cell lineages in the nematode *Caenorhabditis elegans.* In *Development of the Nervous System,* ed. D. Garrod and J. Feldman. Cambridge: Cambridge University Press, pp. 331–345.

Hubel, D. H. 1982. Exploration of the primary visual cortex, 1955–1978. *Nature* 299:515–524.

Hubel, D. H., and T. N. Wiesel. 1963. Receptive fields in striate cortex of very young, visually inexperienced kittens. *J. Neurophysiol.* 26:994–1002.

——— 1965. Binocular interaction in striate cortex of kittens reared with artificial squint. *J. Neurophysiol.* 28:1041–1059.

——— 1979. Brain mechanisms of vision. *Sci. Amer.* 241(3):150–162.

Hubel, D. H., T. N. Wiesel, and E. Bough. 1970. Stereoscopic vision in macaque monkey. *Nature* 225:41–44.

Hubel, D. H., T. N. Wiesel, and S. LeVay. 1977. Plasticity of ocular dominance columns in monkey striate cortex. *Phil. Trans. R. Soc. Lond.,* ser. B, 278: 377–409.

Hulst, J. R., and M. R. Bennett. 1986. Motoneurone survival factor produced by muscle increases in parallel with messenger RNA sequences homologous to βNGF-cDNA. *Dev. Brain Res.* 25:153–156.

Hume, R. I., and D. Purves. 1981. Geometry of neonatal neurones and the regulation of synapse elimination. *Nature* 293:469–471.

——— 1983. Apportionment of the terminals from single preganglionic axons to target neurones in the rabbit ciliary ganglion. *J. Physiol.* (Lond.) 338:259–275.

Huttenlocher, P. R. 1979. Synaptic density in human frontal cortex. Developmental changes and effects of aging. *Brain Res.* 163:195–205.

Huttenlocher, P. R., and C. de Courten. 1987. The development of synapses in striate cortex of man. *Human Neurobiol.* 6:1–9.

Huttenlocher, P. R., C. de Courten, L. J. Garey, and H. Van der Loos. 1982. Synaptogenesis in the human visual cortex—evidence for synapse elimination during normal development. *Neurosci. Letters* 33:247–252.

Hutter, O. F. 1952. Post-tetanic restoration of neuromuscular transmission blocked by d-tubocurarine. *J. Physiol.* (Lond.) 118:216–227.

Huxley, J. 1932. *Problems of Relative Growth.* London: Methuen.

Innocenti, G. M. 1981. Growth and reshaping of axons in the establishment of visual callosal connections. *Science* 212:824–827.

——— 1982. Development of interhemispheric cortical connections. *Neur. Res. Prog. Bull.* 20:532–540.

Innocenti, G. M., S. Clarke, and R. Kraftsik. 1986. Interchange of callosal and association projections in the developing visual cortex. *J. Neurosci.* 6:1384–1409.

Innocenti, G. M., L. Fiore, and R. Caminiti. 1977. Exuberant projection into the corpus callosum from the visual cortex of newborn cats. *Neurosci. Letters* 4:237–242.

Irintchev, A., and A. Wernig. 1987. Muscle damage and repair in voluntarily running mice: Strain and muscle differences. *Cell Tissue Res.* 249:509–521.

Ivy, G. O., R. M. Akers, and H. P. Killackey. 1979. Differential distribution of callosal projection neurons in the neonatal and adult rat. *Brain Res.* 173:532–537.

Ivy, G. O., and H. P. Killackey. 1981. The ontogeny of the distribution of callosal projection neurons in the rat parietal cortex. *J. Comp. Neurol.* 195:367–389.

——— 1982. Ontogenic changes in the projections of neocortical neurons. *J. Neurosci.* 2:735–743.

Jackson, H., and T. N. Parks. 1982. Functional synapse elimination in the developing avian cochlear nucleus with simultaneous reduction in cochlear nerve axon branching. *J. Neurosci.* 2:1736–1743.

Jackson, P. C. 1983. Reduced activity during development delays the normal rearrangement of synapses in the rabbit ciliary ganglion. *J. Physiol.* (Lond.) 345:319–327.

Jacobs, M. S., W. L. McFarland, and P. J. Morgane. 1979. The anatomy of the brain of the bottlenose dolphin (*Tursiops truncatus*). Rhinic lobe (rhinencephalon): The archicortex. *Brain Res. Bull.* 4 (suppl. 1):1–108.

Jacobson, M. 1978. *Developmental Neurobiology.* New York: Plenum.

——— 1980. Clones and compartments in the vertebrate central nervous system. *Trends in Neurosci.* 3:3–5.

Jeanmonod, D., F. L. Rice, and H. Van der Loos. 1981. Mouse somatosensory cortex: Alterations in the barrelfield following receptor injury at different early postnatal ages. *Neuroscience* 6:1503–1535.

Jellies, J., C. M. Loer, and W. B. Kristan, Jr. 1987. Morphological changes in leech Retzius neurons after target contact during embryogenesis. *J. Neurosci.* 7:2618–2629.

Jerison, H. J. 1961. Quantitative analysis of evolution of the brain in mammals. *Science* 133:1012–1014.

——— 1973. *Evolution of the Brain and Intelligence.* New York: Academic Press.

Johnson, D. A. 1988. Regulation of the connections of mammalian parasympathetic neurons by preganglionic innervation. *J. Physiol.* (Lond.) 397:51–62.

Johnson, D. A., and D. Purves. 1981. Post-natal reduction of neural unit size in the rabbit ciliary ganglion. *J. Physiol.* (Lond.) 318:143–159.

——— 1983. Tonic and reflex synaptic activity recorded in ciliary ganglion cells of anaesthetized rabbits. *J. Physiol.* (Lond.) 339:599–613.

Johnson, E. M., Jr., P. D. Gorin, L. D. Brandeis, and J. Pearson. 1980. Dorsal root ganglion neurons are destroyed by exposure *in utero* to maternal antibody to nerve growth factor. *Science* 210:916–918.

Johnson, E. M., Jr., K. M. Rich, and H. K. Yip. 1986. The role of NGF in sensory neurons *in vivo*. *Trends in Neurosci.* 9:33–37.

Johnson, E. M., Jr., M. Taniuchi, and P. S. DiStefano. 1988. Expression and possible function of nerve growth factor receptors on Schwann cell. *Trends in Neurosci.* (in press).

Johnson, E. M., Jr., and K. H. Yip. 1985. Central nervous system and peripheral nerve growth factor provide trophic support critical to mature sensory neuronal survival. *Nature* 314:751–752.

Johnson, J. E., Y.-A. Barde, and H. Thoenen. 1986. The effects of brain-derived neurotrophic factor (BDNF) on the survival of rat retinal neurons in culture. *Soc. Neurosci. Abstr.* 12:1101.

Jones, R., and G. Vrbová. 1970. Effect of muscle activity on denervation hypersensitivity. *J. Physiol.* (Lond.) 210:144–145p.

Jones, S. P., R. M. A. P. Ridge, and A. Rowlerson. 1987. Rat muscle during postnatal development: Evidence in favour of no interconversion between fast- and slow-twitch fibres. *J. Physiol.* (Lond.) 386:395–406.

Kaas, J. H. 1977. Sensory representations in mammals. In *Function and Formation of Neural Systems,* ed. G. S. Stent. Berlin: Dahlem Konferenzen, pp. 65–80.

Kaas, J. H., M. M. Merzenich, and H. P. Killackey. 1983. The reorganization of somatosensory cortex following peripheral nerve damage in adult and developing mammals. *Ann. Rev. Neurosci.* 6:325–356.

Kalcheim, C., Y-A. Barde, H. Thoenen, and N. Le Douarin. 1987. *In vivo* effect of brain-derived neurotrophic factor on the survival of developing dorsal root ganglion cells. *EMBO J.* 6:2871–2873.

Kandel, E. R. 1979a. *Cellular Basis of Behavior: An Introduction to Behavioral Neurobiology*. New York: W. H. Freeman.

——— 1979b. Small systems of neurons. *Sci. Am.* 241(3):60–70.

——— 1979c. Cellular insights into behavior and learning. *Harvey Lect.* 73:19–92.

Kandel, E. R., and J. H. Schwartz. 1982. Molecular biology of learning: Modulation of transmitter release. *Science* 218:433–443.

Kaplan, M. S., and J. W. Hinds. 1977. Neurogenesis in the adult rat: Electron-microscopic analysis of light radioautographs. *Science* 197:1092–1094.

Kashihara, Y., M. Kuno, and Y. Miyata. 1987. Cell death of axotomized motoneurones in neonatal rats, and its prevention by peripheral reinnervation. *J. Physiol.* (Lond.) 386:135–148.

Kater, S. B., and C. Nicholson, eds. 1973. *Intracellular Staining in Neurobiology.* New York: Springer-Verlag.

Katz, B. 1969. *The Release of Neural Transmitter Substances.* Liverpool: Liverpool University Press.

Katz, B., and R. Miledi. 1968. The role of calcium in neuromuscular facilitation. *J. Physiol.* (Lond.) 195:481–492.

Kawamura, Y., and P. J. Dyck. 1981. Permanent axotomy by amputation results in loss of motor neurons in man. *J. Neuropath. and Exp. Neurol.* 40:658–666.

Kelso, S. R., A. H. Ganong, and T. H. Brown. 1986. Hebbian synapses in hippocampus. *Proc. Natl. Acad. Sci.* (U.S.) 83:5326–5330.

Kennard, M. A. 1936. Age and other factors in motor recovery from precentral lesions in monkeys. *Am. J. Physiol.* 115:138–146.

——— 1942. Cortical reorganization of motor function: Studies on series of monkeys of various ages from infancy to maturity. *Arch. Neurol. Psychi.* 48:227–240.

Kimble, J., and D. Hirsh. 1979. The postembryonic cell lineages of the hermaphrodite and male gonads in *Caenorhabditis elegans. Dev. Biol.* 70:396–417.

Kimble, J., J. Sulston, and J. White. 1979. Regulative development in the postembryonic lineages of *Caenorhabditis elegans.* In *Cell Lineage, Stem Cells, and Cell Determination,* INSERM Symp. No. 10, ed. N. LeDouarin. Amsterdam: Elsevier, pp. 59–68.

Kimmel, C. B., and R. W. Warga. 1986. Tissue-specific cell lineages originate in the gastrula of zebrafish. *Science* 231:365–368.

——— 1987a. Cell lineages generating axial muscle in the zebrafish embryo. *Nature* 327:234–237.

——— 1987b. Indeterminate cell lineage of the zebrafish embryo. *Dev. Biol.* 124:269–280.

Kline, M. J., and S. A. Moody. 1987. Fates of the blastomeres of the 4- and 8-cell stage frog (*Xenopus laevis*) embryos. *Anat. Rec.* 218:73A–74A.

Knudsen, E. I. 1980. Sound localization in birds. In *Comparative Studies of Hearing in Vertebrates,* ed. A. N. Popper and R. R. Fay. New York: Springer-Verlag, pp. 287–322.

——— 1985a. Auditory experience influences the development of sound localization and space encoding in the auditory system. In *Comparative Neurobiology,* ed. M. Cohen and F. Strumwasser. New York: Wiley, pp. 93–115.

——— 1985b. Experience alters the spatial tuning of auditory units in the optic tectum during a sensitive period in the barn owl. *J. Neurosci.* 5:3094–3109.

Knudsen, E. I., S. D. Esterly, and P. F. Knudsen. 1984. Monaural occlusion alters sound localization during a sensitive period in the barn owl. *J. Neurosci.* 4:1001–1011.

Knudsen, E. I., P. F. Knudsen, and S. D. Esterly. 1982. Early auditory experience modifies sound localization in barn owls. *Nature* 295:238–240.

——— 1984. A critical period for the recovery of sound localization accuracy following monaural occlusion in the barn owl. *J. Neurosci.* 4:1012–1020.

Knudsen, E. I., and M. Konishi. 1978a. A neural map of auditory space in the owl. *Science* 200:795–797.

——— 1978b. Space and frequency are represented separately in auditory midbrain of the owl. *J. Neurophysiol.* 41:870–884.

——— 1982. A theory of neural auditory space: Auditory representation in the owl and its significance. In *Cortical Sensory Organization*, vol. 3: *Multiple Auditory Areas*, ed. C. N. Woolsey. Clifton, N.J.: Humana, pp. 219–229.

Knudsen, E. I., M. Konishi, and J. D. Pettigrew. 1977. Receptive fields of auditory neurons in the owl. *Science* 198:1278–1280.

Korsching, S. 1986. The role of nerve growth factor in the CNS. *Trends in Neurosci.* 9:570–573.

Korsching, S., G. Auburger, R. Heumann, J. Scott, and H. Thoenen. 1985. Levels of nerve growth factor and its mRNA in the central nervous system of the rat correlate with cholinergic innervation. *EMBO J.* 4:1389–1393.

Korsching, S. I., R. Heumann, A. Davies, and H. Thoenen. 1986. Levels of nerve growth factor and its mRNA during development and regeneration of the peripheral nervous system. *Soc. Neurosci. Abstr.* 12:1096.

Korsching, S., and H. Thoenen. 1983. Nerve growth factor in sympathetic ganglia and corresponding target organs of the rat: Correlation with density of sympathetic innervation. *Proc. Natl. Acad. Sci.* (U.S.) 80:3513–3516.

——— 1985. Treatment with 6-hydroxydopamine and colchicine decreases nerve growth factor levels in sympathetic ganglia and increases them in the corresponding target tissues. *J. Neurosci.* 5:1058–1061.

Krasne, F. B. 1976. Invertebrate systems as a means of gaining insight into the nature of learning and memory. In *Neural Mechanisms of Learning and Memory*, ed. M. R. Rosenzweig and E. L. Bennett. Cambridge, Mass: MIT Press, pp. 401–429.

Kriegstein, A. R., and M. A. Dichter. 1983. Morphological classification of rat cortical neurons in cell culture. *J. Neurosci.* 3:1634–1637.

Kuffler, S. W., M. J. Dennis, and A. J. Harris. 1971. The development of chemosensitivity in extrasynaptic areas of the neuronal surface after denervation of parasympathetic ganglion cells in the heart of the frog. *Proc. R. Soc. Lond.*, ser. B, 177:555–563.

Kuffler, S. W., and D. Yoshikami. 1975. The distribution of acetylcholine sensitivity at the post-synaptic membrane of vertebrate skeletal twitch muscles: Iontophoretic mapping in the micron range. *J. Physiol.* (Lond.) 244:703–730.

Kuno, M., and R. Llinás. 1970. Alterations of synaptic action on chromatolysed motoneurones of the cat. *J. Physiol.* (Lond.) 210:823–838.

Kuno, M., S. A. Turkanis, and J. N. Weakly. 1971. Correlation between nerve terminal size and transmitter release at the neuromuscular junction of the frog. *J. Physiol.* (Lond.) 213:545–556.

Kurz, E. M., D. R. Sengelaub, and A. P. Arnold. 1986. Androgens regulate the dendritic length of mammalian motoneurons in adulthood. *Science* 232: 395–398.

Kyle, H. M. 1921. The asymmetry, metamorphosis and origin of flat-fishes. *Phil. Trans. Roy. Soc. Lond.*, ser. B, 211:75–129.

Laing, N. G., and M. C. Prestige. 1978. Prevention of spontaneous motoneurone death in chick embryos. *J. Physiol.* (Lond.) 282:33–34.

Laiwand, R., R. Werman, and Y. Yarom. 1987. Time course and distribution of motoneuronal loss in the dorsal motor vagal nucleus of guinea pig after cervical vagotomy. *J. Comp. Neurol.* 256:527–537.

LaMantia, A-S., and P. Rakic. 1984. The number, size, myelination, and regional variation of axons in the corpus callosum and anterior commissure of the developing rhesus monkey. *Soc. Neurosci, Abstr.* 10:1081.

Lamb, A. H. 1980. Motoneurone counts in *Xenopus* frogs reared with one bilaterally innervated hindlimb. *Nature* 284:347–350.

——— 1981. Axon regeneration by developing limb motoneurones in *Xenopus laevis*. *Brain Res.* 209:315–323.

Lance-Jones, C. 1982. Motor neuron cell death in the developing lumbar spinal cord of the mouse. *Dev. Brain Res.* 4:473–479.

Land, P. W., and R. D. Lund. 1979. Development of the rat's uncrossed retinotectal pathway and its relation to plasticity studies. *Science* 205:698–700.

Landis, D. M. D. 1988. Changes in the structure of synaptic junctions during climbing fiber synaptogenesis. *Synapse* (in press).

Landmesser, L. 1980. The generation of neuromuscular specificity. *Ann. Rev. Neurosci.* 3:279–302.

Landmesser, L., and G. Pilar. 1974. Synapse formation during embryogenesis on ganglion cells lacking a periphery. *J. Physiol.* (Lond.) 241:715–736.

Langley, J. N. 1892. On the origin from the spinal cord of the cervical and upper thoracic sympathetic fibres, with some observations on white and grey rami communicantes. *Phil. Trans. R. Soc. Lond.*, ser. B, 183:85–124.

——— 1895. Note on regeneration of pre-ganglionic fibres of the sympathetic. *J. Physiol.* (Lond.) 18:280–284.

——— 1897. On the regeneration of pre-ganglionic and post-ganglionic visceral nerve fibres. *J. Physiol.* (Lond.) 22:215–230.

——— 1921. *The Autonomic Nervous System, Part I.* Cambridge: Heffer and Sons.

Lapicque, M. L. 1946. Cytoarchitectonique du ganglion sympathique en fonction du poids du corps. *Comptes Rendus Acad. Sci.* 221:255–258.

Large, T. H., S. C. Bodary, D. O. Clegg, G. Weskamp, U. Otten, and L. F. Reichardt. 1986. Nerve growth factor gene expression in the developing rat brain. *Science* 234:352–355.

Larrabee, M. G., and D. W. Bronk. 1947. Prolonged facilitation of synaptic excitation in sympathetic ganglia. *J. Neurophysiol.* 10:139–154.

Le Gros Clark, W. E. 1959. The crucial evidence for human evolution. *Proc. Amer. Phil. Soc.* 103:159–172.

Letinsky, M. S., K. H. Fishbeck, and U. J. McMahan. 1976. Precision of reinnervation of original postsynaptic sites in frog muscle after nerve crush. *J. Neurocytol.* 5:691–718.

LeVay, S., M. P. Stryker, and C. J. Shatz. 1978. Ocular dominance columns and their development in layer IV of the cat's visual cortex: A quantitative study. *J. Comp. Neurol.* 179:223–244.

LeVay, S., T. N. Wiesel, and D. H. Hubel. 1980. The development of ocular dominance columns in normal and visually deprived monkeys. *J. Comp. Neurol.* 191:1–51.

Levi, G. 1905. Studî sulla grandezza delle cellule. *Archivio Ital. Anat. Embriol.* 5:291–358.

——— 1930. Modificazione dei neuroni simpatici dei mammiferi domestici in relazione all'accrescimento somatico ed alla senescenza. *Boll. Soc. Ital. Biol. Sper.* 5:299–302.

Levi, R., and E. Sacerdote. 1934. Ricerche quantitative sul sistema nervoso di mus musculus. Variazioni nel numero dei neuroni sensitivi spinali in esemplari della stessa famiglia e della stessa specie. *Monitore Zool. Ital.* 45:162–172.

Levi-Montalcini, R. 1949. The development of the acoustico-vestibular centers in the chick embryo in the absence of the afferent root fibers and of descending fiber tracts. *J. Comp. Neurol.* 91:209–241.

——— 1953. Effects of mouse tumor transplantation on the nervous system. *Ann. N.Y. Acad. Sci.* 55:330–343.

——— 1972. The morphological effects of immunosympathectomy. In *Immunosympathectomy,* ed. G. Steiner and E. Schonbaum. Amsterdam: Elsevier, pp. 55–77.

——— 1987. The nerve growth factor: Thirty-five years later. *EMBO J.* 6:1145–1154.

Levi-Montalcini, R., and B. Booker. 1960. Destruction of the sympathetic ganglia in mammals by an antiserum to a nerve growth protein. *Proc. Natl. Acad. Sci.* (U.S.) 46:384–391.

Levi-Montalcini, R., and S. Cohen. 1956. *In vitro* and *in vivo* effects of a nerve growth-stimulating agent isolated from snake venom. *Proc. Natl. Acad. Sci.* (U.S.) 42:695–699.

Levi-Montalcini, R., and V. Hamburger. 1951. Selective growth-stimulating effects of mouse sarcoma on the sensory and sympathetic nervous system of the chick embryo. *J. Exp. Zool.* 116:321–361.

——— 1953. A diffusible agent of mouse sarcoma producing hyperplasia of sympathetic ganglia and hyperneurotization of viscera in the chick embryo. *J. Exp. Zool.* 123:233–287.

Levi-Montalcini, R., and G. Levi. 1942. Les conséquences de la déstruction d'un territoire d'innervation périphérique sur le développement des centres nerveux correspondants dans l'embryon de poulet. *Arch. Biol.* (Liège) 53:537–545.

——— 1944. Correlazioni nello sviluppo tra varie parti del sistema nervoso. *Pontif. Acad. Sci. Commentat.* 8:529–568.

Levi-Montalcini, R., H. Meyer, and V. Hamburger. 1954. *In vitro* experiments on the effects of mouse sarcomas 180 and 37 on the spinal and sympathetic ganglia of the chick embryo. *Cancer Res.* 14:49–57.

Levine, R. B. 1987. Neural reorganization and its endocrine control during insect metamorphosis. *Current Topics in Dev. Biol.* 21:341–365.

Levine, R. B., and J. W. Truman. 1985. Dendritic reorganization of abdominal motoneurons during metamorphosis of the moth, *Manduca sexta*. *J. Neurosci.* 5:2424–2431.

Levine, R. B., J. W. Truman, and C. M. Bate. 1986. Endocrine regulation of the form and function of axonal arbors during insect metamorphosis. *J. Neurosci.* 6:293–299.

Levinthal, R., E. Macagno, and C. Levinthal. 1976. Anatomy and development of identified cells in isogenic organisms. In *Cold Spring Harbor Symp. Quant. Biol.* 40:321–331.

Lichtman, J. W. 1977. The reorganization of synaptic connexions in the rat submandibular ganglion during post-natal development. *J. Physiol.* (Lond.) 273:155–177.

——— 1980. On the predominantly single innervation of submandibular ganglion cells in the rat. *J. Physiol.* (Lond.) 302:121–130.

Lichtman, J. W., L. Magrassi, and D. Purves. 1987. Visualization of neuromuscular junctions over periods of several months in living mice. *J. Neurosci.* 7:1215–1222.

Lichtman, J. W., and D. Purves. 1980. The elimination of redundant preganglionic innervation to hamster sympathetic ganglion cells in early postnatal life. *J. Physiol.* (Lond.) 301:213–228.

——— 1981. Regulation of the number of axons that innervate target cells. In *Development in the Nervous System*, ed. D. R. Garrod and J. D. Feldman. Cambridge: Cambridge University Press, pp. 233–243.

Lichtman, J. W., D. Purves, and J. W. Yip. 1980. Innervation of neurones in the guinea-pig thoracic chain. *J. Physiol.* (Lond.) 298:285–299.

Lichtman, J. W., R. S. Wilkinson, and M. M. Rich. 1985. Multiple innervation of tonic endplates revealed by activity-dependent uptake of fluorescent probes. *Nature* 314:357–359.

Lieberman, A. R. 1971. The axon reaction: A review of the principal features of perikaryal responses to axon injury. *Int. Rev. Neurobiol.* 14:49–124.

Liley, A. W. 1956. An investigation of spontaneous activity at the neuromuscular junction of the rat. *J. Physiol.* (Lond.) 132:650–656.

Lindsay, R. M., H. Thoenen, and Y.-A. Barde. 1985. Placode and neural crest-derived sensory neurons are responsive at early developmental stages to brain-derived neurotrophic factor. *Dev. Biol.* 112:319–328.

Lloyd, D. P. C. 1950. Post-tetanic potentiation of response in monosynaptic reflex pathways of the spinal cord. *J. Gen. Physiol.* 33:147–170.

Loer, C. M., J. Jellies, and W. B. Kristan, Jr. 1987. Segment-specific morphogenesis of leech Retzius neurons requires particular peripheral targets. *J. Neurosci.* 7:2630–2638.

Lømo, T., and J. Rosenthal. 1972. Control of ACh sensitivity by muscle activity in the rat. *J. Physiol.* (Lond.) 221:493–513.

Lømo, T., and R. H. Westgaard. 1975. Further studies on the control of ACh sensitivity by muscle activity in the rat. *J. Physiol.* (Lond.) 252:603–626.

——— 1976. Control of ACh sensitivity in rat muscle fibers. *Cold Spring Harbor Symp. Quant. Biol.* 40:263–274.

Longley, A. 1978. Anatomical mapping of retino-tectal connections in developing and metamorphosed *Xenopus:* Evidence for changing connections. *J. Embryol. Exp. Morphol.* 45:249–270.

Luckenbill-Edds, L., and S. C. Sharma. 1977. Retinotectal projection of the adult winter flounder (*Pseudopleuronectes americanus*). *J. Comp. Neurol.* 173:307–318.

Lund, J. S., R. G. Booth, and R. D. Lund. 1977. Development of neurons in the visual cortex (area 17) of the monkey (*Macaca nemestrina*): A Golgi study from fetal day 127 to postnatal maturity. *J. Comp. Neurol.* 176:149–188.

Lund, R. D., T. J. Cunningham, and J. S. Lund. 1973. Modified optic projections after unilateral eye removal in young rats. *Brain Behav. Evol.* 8:51–72.

Macagno, E. R., A. Peinado, and R. R. Stewart. 1986. Segmental differentiation in the leech nervous system: Specific phenotypic changes associated with ectopic targets. *Proc. Natl. Acad. Sci.* (U.S.) 83:2746–2750.

Mackintosh, N. A., and J. F. G. Wheeler. 1929. Southern blue and fin whales. *"Discovery" Reports, Falkland Islands* 1:257–540.

Maehlen, J., and A. Njå. 1981. Selective synapse formation during sprouting after partial denervation of the guinea-pig superior cervical ganglion. *J. Physiol.* (Lond.) 319:555–567.

——— 1984. Rearrangement of synapses on guinea-pig sympathetic ganglion cells after partial interruption of the preganglionic nerve. *J. Physiol.* (Lond.) 348:43–56.

Magrassi, L., D. Purves, and J. W. Lichtman. 1987. Fluorescent probes that stain living nerve terminals. *J. Neurosci.* 7:1207–1214.

Mallart, A., D. Angaut-Petit, N. F. Zilber-Gachelin, J. Tomás i Ferré, and C. Haimann. 1980. Synaptic efficacy and turnover of endings in pauci-innervated muscle fibres of *Xenopus laevis.* In *Ontogenesis and Functional Mechanisms of Peripheral Synapses,* INSERM Symp. No. 13, ed. J. Taxi. Amsterdam: Elsevier, pp. 213–223.

Manthorpe, M., S. Skaper, R. Adler, K. Landa, and S. Varón. 1980. Cholinergic neuronotrophic factors: Fractionation properties of an extract from selected chick embryonic eye tissues. *J. Neurochem.* 34:69–75.

Mariani, J., and J.-P. Changeux. 1980. Multiple innervation of Purkinje cells by climbing fibers in the cerebellum of the adult *staggerer* mutant mouse. *J. Neurobiol.* 11:41–50.

——— 1981. Ontogenesis of olivocerebellar relationships. I. Studies by intracellular recordings of the multiple innervation of Purkinje cells by climbing fibers in the developing rat cerebellum. *J. Neurosci.* 1:696–702.

Marler, P. 1981. Birdsong: The acquisition of a learned motor skill. *Trends in Neurosci.* 5:88–94.

Matthews, M. R., and V. H. Nelson. 1975. Detachment of structurally intact nerve endings from chromatolytic neurones of the rat superior cervical ganglion during the depression of synaptic transmission induced by post-ganglionic axotomy. *J. Physiol.* (Lond.) 245:91–135.

Matthews, M. R., and T. P. S. Powell. 1962. Some observations on trans-neuronal cell degeneration in the olfactory bulb of the rabbit. *J. Anat.* (Lond.) 96:89–102.

McCully, K. K., and J. A. Faulkner. 1985. Injury to skeletal muscle fibers of mice following lengthening contractions. *J. Appl. Physiol.* 59:119–126.

Mendell, L. M. 1984. Modifiability of spinal synapses. *Physiol. Rev.* 64:260–324.

Mendell, L. M., J. B. Munson, and J. G. Scott. 1976. Alterations of synapses on axotomized motoneurones. *J. Physiol.* (Lond.) 255:67–79.

Merzenich, M. M. 1985. Sources of intraspecies and interspecies cortical map variability in mammals: Conclusions and hypotheses. In *Comparative Neurobiology: Modes of Communication in the Nervous System,* ed. M. J. Cohen and F. Strumwasser. New York: Wiley, pp. 105–116.

Merzenich, M. M., R. J. Nelson, J. H. Kaas, M. P. Stryker, W. M. Jenkins, J. M. Zook, M. S. Cynader, and A. Schoppmann. 1987. Variability in hand surface representations in areas 3b and 1 in adult owl and squirrel monkeys. *J. Comp. Neurol.* 258:281–296.

Meyer, R. L. 1978. Evidence from thymidine labeling for continuing growth of retina and tectum in juvenile goldfish. *Exp. Neurol.* 59:99–111.

Miledi, R. 1960a. The acetylcholine sensitivity of frog muscle fibres after complete or partial denervation. *J. Physiol.* (Lond.) 151:1–23.

——— 1960b. Junctional and extra-junctional acetylcholine receptors in skeletal muscle fibres. *J. Physiol.* (Lond.) 151:24–30.

Milner, P. M. 1986. Obituary: Donald Olding Hebb (1904–1985). *Trends in Neurosci.* 9:347–351.

Moody, S. A. 1987a. Fates of the blastomeres of the 16-cell stage *Xenopus* embryo. *Dev. Biol.* 119:560–578.

——— 1987b. Fates of the blastomeres of the 12-cell stage *Xenopus* embryo. *Dev. Biol.* 122:300–319.

Morgan, D. L., and U. Proske. 1984. Vertebrate slow muscle: Its structure, pattern of innervation, and mechanical properties. *Physiol. Rev.* 64:103–169.

Morgane, P. J., and M. S. Jacobs. 1972. Comparative anatomy of the cetacean nervous system. In *Functional Anatomy of Marine Mammals,* ed. R. J. Harrison. New York: Academic Press, pp. 117–244.

Munro, H. N., ed. 1969. *Mammalian Protein Metabolism,* vol. 3. New York: Academic Press.

Murphey, R. K. 1986a. Competition and the dynamics of axon arbor growth in the cricket. *J. Comp. Neurol.* 251:100–110.

——— 1986b. The myth of the inflexible invertebrate: Competition and synaptic remodelling in the development of invertebrate nervous systems. *J. Neurobiol.* 17:585–591.

Murray, J. G., and J. W. Thompson. 1957. The occurrence and function of collateral sprouting in the sympathetic nervous system of the cat. *J. Physiol.* (Lond.) 135:133–162.

Nishi, R., and D. K. Berg. 1981. Two components from eye tissue that differentially stimulate the growth and development of ciliary ganglion neurons in cell culture. *J. Neurosci.* 1:505–513.

Njå, A., and D. Purves. 1977a. Specific innervation of guinea-pig superior cervical ganglion cells by preganglionic fibres arising from different levels of the spinal cord. *J. Physiol.* (Lond.) 264:565–583.

——— 1977b. Re-innervation of guinea-pig superior cervical ganglion cells by preganglionic fibres arising from different levels of the spinal cord. *J. Physiol.* (Lond.) 272:633–651.

——— 1978a. The effects of nerve growth factor and its antiserum on synapses in the superior cervical ganglion of the guinea-pig. *J. Physiol.* (Lond.) 277:53–75.

——— 1978b. Specificity of initial synaptic contacts made on guinea-pig superior cervical ganglion cells during regeneration of the cervical sympathetic trunk. *J. Physiol.* (Lond.) 281:45–62.

Nordeen, E. J., K. W. Nordeen, D. R. Sengelaub, and A. P. Arnold. 1985. Androgens prevent normally occurring cell death in a sexually dimorphic spinal nucleus. *Science* 229:671–673.

Nornes, H. O., and G. D. Das. 1974. Temporal pattern of neurogenesis in spinal cord of rat. I. An autoradiographic study—time and sites of origin and migration and settling patterns of neuroblasts. *Brain Res.* 73:121–138.

Nottebohm, F. 1980. Testosterone triggers growth of brain vocal control nuclei in adult female canaries. *Brain Res.* 189:429–436.

——— 1981. A brain for all seasons: Cyclical anatomical changes in song control nuclei of the canary brain. *Science* 214:1368–1370.

——— 1984. Birdsong as a model in which to study brain processes related to learning. *Condor* 86:227–236.

Nottebohm, F., S. Kasparian, and C. Pandazis. 1981. Brain space for a learned task. *Brain Res.* 213:99–109.

Nottebohm, F., and M. E. Nottebohm. 1978. Relationship between song repertoire and age in the canary, *Serinus canarius*. *Z. Tierpsychol.* 46:298–305.

Nudell, B. M., and A. D. Grinnell. 1983. Regulation of synaptic position, size, and strength in anuran skeletal muscle. *J. Neurosci.* 3:161–176.

Nurcombe, V., M. A. Hill, K. L. Eagleson, and M. R. Bennett. 1984. Motoneurone survival and neuritic extension from spinal cord explants induced by factors released from denervated muscle. *Brain Res.* 291:19–28.

Nyström, B. 1968a. Histochemical studies of endplate bound esterases in "slow-red" and "fast-white" cat muscles during postnatal development. *Acta Neurol. Scand.* 44:295–318.

——— 1968b. Postnatal development of motor nerve terminals in "slow-red" and "fast-white" cat muscles. *Acta Neurol. Scand.* 44:363–383.

O'Brien, R. A. D., A. J. C. Ostberg, and G. Vrbová. 1978. Observations on the elimination of polyneuronal innervation in developing mammalian skeletal muscle. *J. Physiol.* (Lond.) 282:571–582.

Ojemann, G. A. 1983. Brain organization for language from the perspective of electrical stimulation mapping. *Behav. Brain Sci.* 2:189–230.

Okado, N., and R. W. Oppenheim. 1984. Cell death of motoneurons in the chick embryo spinal cord: IX. The loss of motoneurons following removal of afferent inputs. *J. Neurosci.* 4:1639–1652.

O'Kusky, J., and M. Colonnier. 1982. Postnatal changes in number of neurons and synapses in visual cortex (area 17) of macaque monkey: A stereological analysis in normal and monocularly deprived animals. *J. Comp. Neurol.* 210:291–306.

O'Leary, D. D. M. 1987. Remodelling of early axonal projections through the selective elimination of neurons and long axon collaterals. In *Selective Neuronal Death,* ed. M. O'Connor. *CIBA Found. Symp.* 126:113–142.

O'Leary, D. D. M., J. W. Fawcett, and W. M. Cowan. 1986. Topographic targeting errors in the retinocollicular projection and their elimination by selective ganglion cell death. *J. Neurosci.* 6:3692–3705.

O'Leary, D. D. M., C. R. Gerfen, and W. M. Cowan. 1983. The development and restriction of the ipsilateral retinofugal projection in the chick. *Dev. Brain Res.* 10:93–109.

O'Leary, D. D. M., and B. B. Stanfield. 1986. A transient pyramidal tract projection from the visual cortex in the hamster and its removal by selective collateral elimination. *Dev. Brain Res.* 27:87–99.

Olson, L. 1967. Outgrowth of sympathetic adrenergic neurons in mice treated with a nerve growth factor (NGF). *Z. Zellforsch. mikros. Anat.* 81: 155–173.

Olson, L., and T. Malmfors. 1970. Growth characteristics of adrenergic nerves in the adult rat. Fluorescence, histochemical, and ^{3}H-noradrenaline uptake studies using tissue transplantations to the anterior chamber of the eye. *Acta Physiol. Scand.,* suppl. 348:1–111.

Ontell, M., and R. F. Dunn. 1978. Neonatal muscle growth: A quantitative study. *Am. J. Anat.* 152:539–556.

Oppenheim, R. W. 1981. Neuronal cell death and some related regressive phenomena during neurogenesis: A selective historical review and a progress report. In *Studies in Developmental Neurobiology: Essays in Honor of Viktor Hamburger,* ed. W. M. Cowan. New York: Oxford University Press, pp. 74–133.

——— 1985. Naturally occurring cell death during neural development. *Trends in Neurosci.* 8:487–493.

——— 1986. The absence of significant postnatal motor neuron death in the brachial and lumbar spinal cord of the rat. *J. Comp. Neurol.* 246:281–286.

Oppenheim, R. W., I.-W. Chu-Wang, and J. L. Maderdrut. 1978. Cell death of motoneurons in the chick embryo spinal cord. III. The differentiation of motoneurons prior to their induced degeneration following limbbud removal. *J. Comp. Neurol.* 177:87–111.

Oppenheim, R. W., L. J. Haverkamp, D. Prevette, and S. Appel. 1987. Reduction of naturally occurring motoneuron death *in vivo* by a putative target-derived neurotrophic factor. *Soc. Neurosci. Abstr.* 13:1014.

O'Rourke, N. A., and S. E. Fraser. 1986. Dynamic aspects of retinotectal map formation revealed by a vital-dye fiber-tracing technique. *Dev. Biol.* 114:265–276.

Owman, C. 1981. Pregnancy induces degenerative and regenerative changes in the autonomic innervation of the female reproductive tract. In *Development of the Autonomic Nervous System*, CIBA Foundation Symp. 83, ed. K. Elliott and G. Lawrenson. London: Pittman, pp. 252–279.

Pakkenberg, H., and J. Voigt. 1964. Brain weight of the Danes. *Acta Anat.* 56:297–307.

Palay, S. L., and V. Chan-Palay. 1974. *Cerebellar Cortex.* New York: Springer-Verlag.

——— eds. 1982. *The Cerebellum. New Vistas. Exp. Brain Res.*, suppl. 6.

Palmitier, M. A., B. K. Hartman, and E. M. Johnson, Jr. 1984. Demonstration of retrogradely transported endogenous nerve growth factor in axons of sympathetic neurons. *J. Neurosci.* 4:751–756.

Parks, T. N. 1979. Afferent influences on the development of the brain stem auditory nuclei of the chicken: Otocyst ablation. *J. Comp. Neurol.* 183:665–678.

Pearce, J., C. K. Govind, and D. E. Meiss. 1985. Growth-related features of lobster neuromuscular terminals. *Dev. Brain Res.* 21:215–228.

Pearson, K. G., and C. S. Goodman. 1979. Correlation of variability in structure with variability in synaptic connections of an identified interneuron in locusts. *J. Comp. Neurol.* 184:141–165.

Peichl, L., H. Ott, and B. B. Boycott. 1987. Alpha-ganglion cells in mammalian retinae. *Proc. Roy. Soc. Lond.*, ser. B, 231:169–197.

Penfield, W., and E. Boldrey. 1937. Somatic motor and sensory representation in the cerebral cortex of man as studied by electrical stimulation. *Brain* 60:389–443.

Penfield, W., and H. Jasper. 1954. *Epilepsy and the Functional Anatomy of the Human Brain.* Boston: Little, Brown.

Perry, G. W., S. R. Krayanek, and D. L. Wilson. 1983. Protein synthesis and rapid axonal transport during regrowth of dorsal root axons. *J. Neurochem.* 40:1590–1598.

Pestronk, A., and D. B. Drachman. 1978. Motor nerve sprouting and acetylcholine receptors. *Science* 199:1223–1225.

——— 1985. Motor nerve terminal outgrowth and acetylcholine receptors: Inhibition of terminal outgrowth by α-bungarotoxin and anti-acetylcholine receptor antibody. *J. Neurosci.* 5:751–758.

Peusner, K. D., and D. K. Morest. 1977. Neurogenesis in the nucleus vestibularis tangentialis of the chick embryo in the absence of the primary afferent fibers. *Neurosci.* 2:253–270.

Phillips, B. F., J. S. Cobb, and R. W. George. 1980. General biology. In *The Biology and Management of Lobsters*, vol. 1, ed. J. S. Cobb and B. F. Phillips. New York: Academic Press, pp. 1–89.

Pick, J. 1970. *The Autonomic Nervous System*. Philadelphia: Lippincott.

Pilar, G., and L. Landmesser. 1972. Axotomy mimicked by localized colchicine application. *Science* 177:1116–1118.

——— 1976. Ultrastructural differences during embryonic cell death in normal and peripherally deprived ciliary ganglia. *J. Cell Biol.* 68:339–356.

Pilar, G., L. Landmesser, and L. Burstein. 1980. Competition for survival among developing ciliary ganglion cells. *J. Neurophysiol.* 43:233–254.

Pittman, R., and R. W. Oppenheim. 1979. Cell death of motoneurons in the chick embryo spinal cord. IV. Evidence that a functional neuromuscular interaction is involved in the regulation of naturally occurring cell death and the stabilization of synapses. *J. Comp. Neurol.* 187:425–446.

Platt, C. 1973a. Central control of postural orientation in flatfish. I. Postural change dependence on central neural changes. *J. Exp. Biol.* 59:491–521.

——— 1973b. Central control of postural orientation in flatfish. II. Optic-vestibular efferent modification of gravistatic input. *J. Exp. Biol.* 59:523–541.

Polyak, S. 1926. The connections of the acoustic nerve. *J. Anat.* 60:465–469.

Powell, T. P. S., and S. D. Erulkar. 1962. Transneuronal cell degeneration in the auditory relay nuclei of the cat. *J. Anat.* 91:249–268.

Purves, D. 1975. Functional and structural changes in mammalian neurones following interruption of their axons. *J. Physiol.* (Lond.) 252:429–463.

——— 1976a. Functional and structural changes in mammalian sympathetic neurones following colchicine application to postganglionic nerves. *J. Physiol.* (Lond.) 259:159–175.

——— 1976b. Long-term regulation in the vertebrate peripheral nervous system. In *International Review of Physiology, Neurophysiology II*, vol. 10, ed. R. Porter. Baltimore: University Park Press, pp. 125–177.

——— 1977. The formation and maintenance of synaptic connections. In *Function and Formation of Neural Systems*, ed. G. S. Stent. Berlin: Dahlem Konferenzen, pp. 21–49.

——— 1980. Neuronal competition. *Nature* 287:585–586.

——— 1986. The trophic theory of neural connections. *Trends in Neurosci.* 9:486–489.

Purves, D., R. D. Hadley, and J. T. Voyvodic. 1986. Dynamic changes in the dendritic geometry of individual neurons visualized over periods of up to three months in the superior cervical ganglion of living mice. *J. Neurosci.* 6:1051–1060.

Purves, D., and R. I. Hume. 1981. The relation of postsynaptic geometry to the number of presynaptic neurones that innervate autonomic ganglion cells. *J. Neurosci.* 1:441–452.

Purves, D., and J. W. Lichtman. 1978. The formation and maintenance of synaptic connections in autonomic ganglia. *Physiol. Rev.* 58:821–862.

——— 1980. Elimination of synapses in the developing nervous system. *Science* 210:153–157.

——— 1985a. Geometrical differences among homologous neurons in mammals. *Science* 228:298–302.

——— 1985b. *Principles of Neural Development.* Sunderland, Mass.: A. Sinauer Associates.

——— 1987. Synaptic sites on reinnervated nerve cells visualized at two different times in living mice. *J. Neurosci.* 7:1492–1497.

Purves, D., and A. Njå. 1978. Trophic maintenance of synaptic connections in autonomic ganglia. In *Neuronal Plasticity,* ed. C. W. Cotman. New York: Raven Press, pp. 27–49.

Purves, D., E. Rubin, W. D. Snider, and J. W. Lichtman. 1986. The relation of animal size to convergence, divergence, and neuronal number in peripheral sympathetic pathways. *J. Neurosci.* 6:153–163.

Purves, D., W. J. Thompson, and J. W. Yip. 1981. Reinnervation of ganglia transplanted to the neck from different levels of the guinea pig sympathetic chain. *J. Physiol.* (Lond.) 313:49–63.

Purves, D., and J. T. Voyvodic. 1987. Imaging mammalian nerve cells and their connections over time in living animals. *Trends in Neurosci.* 10:398–403.

Purves, D., J. T. Voyvodic, L. Magrassi, and H. Yawo. 1987. Nerve terminal remodeling visualized in living mice by repeated examination of the same neuron. *Science* 238:1122–1126.

Purves, D., and D. J. Wigston. 1983. Neural units in the superior cervical ganglion of the guinea-pig. *J. Physiol.* (Lond.) 334:169–178.

Quinn, W. G. 1984. Work in invertebrates on the mechanisms underlying learning. In *The Biology of Learning,* ed. P. Marler and H. S. Terrace. Berlin: Springer-Verlag, pp. 197–246.

Rager, G. 1983. Structural analysis of fiber organization during development. *Prog. Brain Res.* 58:313–319.

Rakic, P. 1974. Neurons in rhesus monkey visual cortex: Systematic relation between time of origin and eventual disposition. *Science* 183:425–427.

——— 1977. Prenatal development in the visual system in the rhesus monkey. *Phil. Trans. R. Soc. Lond.,* ser. B, 278:245–260.

——— 1981. Development of visual centers in the primate brain depends on binocular competition before birth. *Science* 214:928–931.

——— 1985a. DNA synthesis and cell division in the adult primate brain. *Ann. N. Y. Acad. Sci.* 457:193–211.

——— 1985b. Limits of neurogenesis in primates. *Science* 227:1054–1056.

Rakic, P., J.-P. Bourgeois, M. E. Eckenhoff, N. Zečevic, and P. S. Goldman-Rakic. 1986. Concurrent overproduction of synapses in diverse regions of the primate cerebral cortex. *Science* 232:232–235.

Rakic, P., and R. S. Nowakowski. 1981. The time of origin of neurons in the hippocampal region of the rhesus monkey. *J. Comp. Neurol.* 196:99–128.

Rakic, P., and K. P. Riley. 1983a. Overproduction and elimination of retinal axons in fetal rhesus monkey. *Science* 219:1441–1444.

——— 1983b. Regulation of axon numbers in the primate optic nerve by prenatal binocular competition. *Nature* 305:135–137.

Rakic, P., and R. W. Williams. 1986. Thalamic regulation of cortical parcellation: an experimental perturbation of the striate cortex in rhesus monkeys. *Soc. Neurosci. Abstr.* 12:1499.

Rall, W. 1959. Branching dendritic trees and motoneuron membrane resistivity. *Exp. Neurol.* 1:491–527.

Ramón y Cajal, S. 1911. *Histologie du système nerveux de l'homme et des vertébrés,* vol. 1. Paris: A. Maloine. Reprint, Madrid: Consejo Superior de Investigaciones Cientificas, Instituto Ramón y Cajal, 1955.

——— 1925. Quelques remarques sur les plaques motrices de la langue des mammifères. *Trav. Lab. Rech. Biol.* 23.

——— 1929. *Studies on Vertebrate Neurogenesis.* (Translation of the 1929 edition by L. Guth.) Springfield, Ill.: Thomas.

Redman, S. J. 1986. Monosynaptic transmission in the spinal cord. *News in Physiol. Sci.* 1:171–174.

Reh, T. A., and M. Constantine-Paton. 1984. Retinal ganglion cell terminals change their projection sites during larval development of *Rana pipiens. J. Neurosci.* 4:442–457.

Reh, T., and K. Kalil. 1982. Development of the pyramidal tract in the hamster. II. An electron microscopic study. *J. Comp. Neurol.* 205:77–88.

Ribchester, R. R., and T. Taxt. 1983. Motor unit size and synaptic competition in rat lumbrical muscles reinnervated by active and inactive motor axons. *J. Physiol.* (Lond.) 344:89–111.

Rich, M. M., and J. W. Lichtman. 1986. Remodeling of endplate sites during reinnervation in the living mouse. *Neurosci. Abstr.* 12:39.

Riley, D. A. 1978. Tenotomy delays the postnatal development of the motor innervation of the rat soleus. *Brain Res.* 143:162–167.

Rockel, A. J., R. W. Hiorns, and T. P. S. Powell. 1980. The basic uniformity in structure of the neocortex. *Brain* 103:221–244.

Rodin, B. E., S. L. Sampagna, and L. Kruger. 1983. An examination of intraspinal sprouting in dorsal root axons with the tracer horseradish peroxidase. *J. Comp. Neurol.* 215:187–198.

Romanes, G. J. 1895. *Darwin and After Darwin.* London: Open Court Publishing Co.

——— 1946. Motor localization and the effects of nerve injury on ventral horn cells of the spinal cord. *J. Anat.* 80:117–131.

Rootman, D. S., W. G. Tatton, and M. Hay. 1981. Postnatal histogenetic death of rat forelimb motoneurons. *J. Comp. Neurol.* 199:17–27.

Roper, S. 1976. The acetylcholine sensitivity of the surface membrane of multiply-innervated parasympathetic ganglion cells in the mudpuppy before and after partial denervation. *J. Physiol.* (Lond.) 254:455–473.

Roper, S., and C.-P. Ko. 1978. Synaptic remodeling in the partially denervated parasympathetic ganglion in the heart of the frog. In *Neuronal Plasticity,* ed. C. W. Cotman. New York: Raven Press, pp. 1–25.

Rotshenker, S., and U. J. McMahan. 1976. Altered patterns of innervation in frog muscle after denervation. *J. Neurocytol.* 5:719–730.

Rubin, E. 1985a. Development of the rat superior cervical ganglion: Ganglion cell aggregation and maturation. *J. Neurosci.* 5:673–684.

——— 1985b. Development of the rat superior cervical ganglion: Ingrowth of preganglionic axons. *J. Neurosci.* 5:685–696.

——— 1985c. Development of the rat superior cervical ganglion: Initial stages of synapse formation. *J. Neurosci.* 5:697–704.

Rush, R. A. 1984. Immunohistochemical localization of endogenous nerve growth factor. *Nature* 312:364–367.

Salminen, A., and V. Vihko. 1983. Susceptibility of mouse skeletal muscles to exercise injuries. *Muscle and Nerve* 6:596–601.

Sanes, J. R. 1983. Roles of extracellular matrix in neural development. *Ann. Rev. Physiol.* 45:581–600.

Sanes, J. R., and J. Covault. 1985. Axon guidance during reinnervation of skeletal muscle. *Trends in Neurosci.* 8:523–528.

Sanes, J. R., and Z. W. Hall. 1979. Antibodies that bind specifically to synaptic sites on muscle fiber basal lamina. *J. Cell Biol.* 83:357–370.

Sanes, J. R., J. G. Hildebrand, and D. J. Prescott. 1976. Differentiation of insect sensory neurons in the absence of their normal synaptic targets. *Dev. Biol.* 52:121–127.

Sanes, J. R., L. M. Marshall, and U. J. McMahan. 1978. Reinnervation of muscle fiber basal lamina after removal of myofibers. *J. Cell Biol.* 78:176–198.

——— 1980. Reinnervation of skeletal muscle: Restoration of the normal synaptic pattern. In *Nerve Repair and Regeneration: Its Clinical and Experimental Basis,* ed. D. L. Jewett and H. R. McCarroll. St. Louis: Mosby, pp. 130–138.

Sargent, P. B. 1983a. Changes in the innervation of cardiac ganglion cells accompanying neuronal growth in post-metamorphic *Xenopus laevis*. *Soc. Neurosci. Abstr.* 9:935.

——— 1983b. The number of synaptic boutons terminating upon *Xenopus* cardiac ganglion cells is directly correlated with cell size. *J. Physiol.* 343:85–104.

——— 1986. Relationship between synaptic size and target cell size in the parasympathetic cardiac ganglion of *Xenopus laevis*. *Soc. Neurosci. Abstr.* 12:1502.

Sargent, P. B., and M. J. Dennis. 1977. Formation of synapses between parasympathetic neurones deprived of preganglionic innervation. *Nature* 268:456–458.

——— 1981. The influence of normal innervation upon abnormal synaptic connections between frog parasympathetic neurons. *Dev. Biol.* 81:65–73.

Sassoon, D., and D. Kelley. 1986. The sexually dimorphic larynx of *Xenopus laevis:* Development and androgen regulation. *Am. J. Anat.* 177:457–472.

Schäfer, T., M. E. Schwab, and H. Thoenen. 1983. Increased formation of preganglionic synapses and axons due to a retrograde trans-synaptic action of nerve growth factor in the rat sympathetic nervous system. *J. Neurosci.* 3:1501–1510.

Schmalbruch, H. 1984. Motoneuron death after sciatic nerve section in newborn rats. *J. Comp. Neurol.* 224:252–258.

Schmidt, J. T., C. M. Cicerone, and S. S. Easter. 1978. Expansion of the half retinal projection to the tectum in goldfish: An electrophysiological and anatomical study. *J. Comp. Neurol.* 177:257–278.

Schmidt-Nielsen, K. 1984. *Scaling: Why Is Animal Size So Important?* Cambridge: Cambridge University Press.

Schneider, G. E. 1973. Early lesions of the superior colliculus: Factors affecting the formation of abnormal retinal projections. *Brain Behav. Evol.* 8:73–109.

Schuetze, S. M., and L. W. Role. 1987. Developmental regulation of nicotinic acetylcholine receptors. *Ann. Rev. Neurosci.* 10:403–457.

Schwartz, J. H., and S. M. Greenberg. 1987. Molecular mechanisms for memory: Second messenger induced modifications of protein kinases in nerve cells. *Ann. Rev. Neurosci.* 10:459–476.

Schwob, J., and J. Price. 1984. The development of axonal connections in the central olfactory system of rats. *J. Comp. Neurol.* 223:177–202.

Seiler, M., and M. E. Schwab. 1984. Specific retrograde transport of nerve growth factor (NGF) from neocortex to nucleus basalis in the rat. *Brain Res.* 300:33–39.

Sengelaub, D. R., and A. P. Arnold. 1986. Development and loss of early projections in a sexually dimorphic rat spinal nucleus. *J. Neurosci.* 6:1613–1620.

Shatz, C. J. 1983. Prenatal development of the cat's retinogeniculate pathway. *J. Neurosci.* 3:482–499.

Sheard, P., and M. Jacobson. 1987. Clonal restriction boundaries in *Xenopus* embryos shown with two intracellular lineage tracers. *Science* 236:851–854.

Shelton, D. L., and L. F. Reichardt. 1984. Expression of the β-nerve growth factor gene correlates with the density of sympathetic innervation in effector organs. *Proc. Natl. Acad. Sci.* (U.S.) 81:7951–7955.

——— 1986. Studies on the expression of the beta nerve growth factor (NGF) gene in the central nervous system: Level and regional distribution of NGF mRNA suggest that NGF functions as a trophic factor for several distinct populations of neurons. *Proc. Natl. Acad. Sci.* (U.S.) 83:2714–2718.

Shepherd, D., and R. K. Murphey. 1986. Competition regulates the efficacy of an identified synapse in crickets. *J. Neurosci.* 6:3152–3160.

Shorey, M. L. 1909. The effect of the destruction of peripheral areas on the differentiation of the neuroblasts. *J. Exp. Zool.* 7:25–63.

Sidman, R. L., and P. Rakic. 1982. Development of the human central nervous system. In *Histology and Histopathology of the Nervous System*, ed. W. Haymaker and R. D. Adams. Springfield, Ill.: Thomas, pp. 3–145.

Sittig, O. 1925. A clinical study of sensory Jacksonian fits. *Brain* 48:233–254.

Small, G. L. 1971. *The Blue Whale.* New York: Columbia University Press.

Smith, M. A., Y.-M. M. Yao, N. E. Reist, C. Magill, B. G. Wallace, and U. J.

McMahan. 1987. Identification of agrin in electric organ extracts and localization of agrin-like molecules in muscle and central nervous system. *J. Expl. Biol.* 132:223–230.

Smith, R. G., J. McManaman, and S. H. Appel. 1985. Trophic effects of skeletal muscle extracts on ventral spinal cord neurons *in vitro:* Separation of a protein with morphologic activity from proteins with cholinergic activity. *J. Cell Biol.* 101:1608–1621.

Smith, R. G., K. Vaca, J. McManaman, and S. H. Appel. 1986. Selective effects of skeletal muscle extract fractions on motoneuron development *in vitro. J. Neurosci.* 6:439–447.

Smith, S. J. 1987. Progress on LTP at hippocampal synapses: A post-synaptic calcium trigger for memory storage? *Trends in Neurosci.* 10:142–144.

Smith, Z. D. J., L. Gray, and E. W. Rubel. 1983. Afferent influences on brain stem auditory nuclei of the chicken: *N. laminaris* dendritic length following monaural conductive hearing loss. *J. Comp. Neurol.* 220:199–205.

Smolen, A. 1981. Postnatal development of ganglionic neurons in the absence of preganglionic input: Morphological synapse formation. *Dev. Brain Res.* 1:49–58.

Snider, W. D. 1987. The dendritic complexity and innervation of submandibular neurons in five species of mammals. *J. Neurosci.* 7:1760–1768.

——— 1988. Nerve growth factor promotes dendritic arborization of sympathetic ganglion cells in developing mammals. *J. Neurosci.* (in press).

Sotelo, C., and S. L. Palay. 1971. Altered axons and axon terminals in the lateral vestibular nucleus of the rat. Possible example of axonal remodeling. *Lab. Invest.* 25:653–671.

Sperry, D. G. 1981. Fiber type composition and postmetamorphic growth of anuran hindlimb muscles. *J. Morphol.* 170:321–345.

Sperry, R. W. 1943a. Effect of 180-degree rotation of the retinal field on visuomotor coordination. *J. Exp. Zool.* 92:263–279.

——— 1943b. Visuomotor coordination in the newt (*Triturus viridescens*) after regeneration of the optic nerve. *J. Comp. Neurol.* 79:33–55.

——— 1956. The eye and the brain. *Sci. Am.* 194(5):48–52.

——— 1963. Chemoaffinity in the orderly growth of nerve fiber patterns and connections. *Proc. Natl. Acad. Sci.* (U.S.) 50:703–710.

——— 1967. Summation. In *The Anatomy of Memory*, ed. D. P. Kimble. Palo Alto: Science and Behavior Books.

Sretavan, D. W., and C. J. Shatz. 1986. Prenatal development of retinal ganglion cell axons: Segregation into eye-specific layers within the cat's lateral geniculate nucleus. *J. Neurosci.* 6:234–251.

Srihari, T., and G. Vrbová. 1978. The role of muscle activity in the differentiation of neuromuscular junctions in slow and fast chick muscles. *J. Neurocytol.* 7:529–540.

Standler, N. A., and J. J. Bernstein. 1982. Degeneration and regeneration of motoneuron dendrites after ventral root crush: Computer reconstruction of dendritic fields. *Exp. Neurol.* 75:600–615.

Stanfield, B. B., and D. D. M. O'Leary. 1985. The transient corticospinal projection from the occipital cortex during the postnatal development of the rat. *J. Comp. Neurol.* 238:236–248.

Steinbach, J. H. 1981. Developmental changes in acetylcholine receptor aggregates at rat skeletal neuromuscular junctions. *Dev. Biol.* 84:267–276.

Steinbach, J. H., and R. J. Bloch. 1986. The distribution of acetylcholine receptors on vertebrate skeletal muscle cells. In *Receptors in Cellular Recognition and Developmental Processes,* ed. R. M. Gorczynski. New York: Academic Press, pp. 183–213.

Stent, G. S. 1973. A physiological mechanism for Hebb's postulate of learning. *Proc. Natl. Acad. Sci.* (U.S.) 70:997–1001.

Stent, G. S., and D. A. Weisblat. 1985. Cell lineage in the development of invertebrate nervous systems. *Ann. Rev. Neurosci.* 8:45–70.

Stephenson, R. S. 1979. Axon reflexes in axolotl limbs: Evidence that branched motor axons reinnervate muscles selectively. *Exp. Neurol.* 64:174–189.

Sternberg, P. W., and H. R. Horvitz. 1986. Pattern formation during vulval development in *C. elegans. Cell* 74:761–772.

Stone, L. S. 1941. Transplantation of the vertebrate eye and return of vision. *Trans. N.Y. Acad. Sci.* 3:208–212.

——— 1944. Functional polarization in the retinal development and its reestablishment in regenerating retinae of rotated grafted eyes. *Proc. Soc. Exp. Biol. Med.* 57:13–14.

Stone, L. S., and L. S. Farthing. 1942. Return of vision four times in the same adult salamander eye (*Triturus viridescens*) repeatedly transplanted. *J. Exp. Zool.* 91:265–285.

Stone, L. S., and I. S. Zaur. 1940. Reimplantation and transplantation of adult eyes in the salamander (*Triturus viridescens*) with return of vision. *J. Exp. Zool.* 85:243–269.

Streichert, L. C., C. Magill, and P. B. Sargent. 1987. Ultrastructural characterization of synaptic boutons in the cardiac ganglia of postmetamorphic and adult *Xenopus laevis. Soc. Neurosci. Abstr.* 13:1425.

Stretton, A. O. W., and E. A. Kravitz. 1968. Neuronal geometry: Determination with a technique of intracellular dye injection. *Science* 162:132–134.

——— 1973. Intracellular dye injection: The selection of procion yellow and its application in preliminary studies of neuronal geometry in the lobster nervous system. In *Intracellular Staining in Neurobiology,* ed. S. B. Kater and C. A. Nicholson. New York: Springer-Verlag, pp. 21–40.

Suga, N. 1984. The extent to which biosonar information is represented by the auditory cortex of the mustached bat. In *Dynamic Aspects of Neocortical Function,* ed. G. M. Edelman, W. E. Gall, and W. M. Cowan. New York: Wiley, pp. 315–373.

Sulston, J. E., and H. R. Horvitz. 1977. Post-embryonic cell lineages of the nematode *Caenorhabditis elegans. Dev. Biol.* 56:110–156.

——— 1981. Abnormal cell lineages in mutants of the nematode *Caenorhabditis elegans. Dev. Biol.* 82:41–55.

Sulston, J. E., E. Schierenberg, J. G. White, and J. N. Thomson. 1983. The embryonic cell lineage of the nematode *Caenorhabditis elegans. Dev. Biol.* 100:64–119.

Sulston, J. E., and J. G. White. 1980. Regulation and cell autonomy during postembryonic development of *Caenorhabditis elegans*. *Dev. Biol.* 78:577–597.

Sumner, B. E. H. 1975. A quantitative analysis of boutons with different types of synapse in normal and injured hypoglossal nuclei. *Exp. Neurol.* 49:406–417.

——— 1976. Quantitative ultrastructural observations on the inhibited recovery of the hypoglossal nucleus from the axotomy response when regeneration of the hypoglossal nerve is prevented. *Exp. Brain Res.* 26:141–150.

Sumner, B. E. H., and F. I. Sutherland. 1973. Quantitative electron microscopy on the injured hypoglossal nucleus in the rat. *J. Neurocytol.* 2:315–328.

Sumner, B. E. H., and W. E. Watson. 1971. Retraction and expansion of the dendritic tree of motor neurones of adult rats induced *in vivo*. *Nature* 233:273–275.

Swanson, L. W., T. J. Teyler, and R. F. Thompson, eds. 1982. Hippocampal long-term potentiation: Mechanisms and implications for memory. *Neurosci. Res. Prog. Bull.* 20:613–769.

Taghert, P. H., C. Q. Doe, and C. S. Goodman. 1984. Cell determination and regulation during development of neuroblasts and neurones in grasshopper embryo. *Nature* 307:163–165.

Tanaka, H., and L. T. Landmesser. 1986. Cell death of lumbosacral motoneurons in chick, quail, and chick-quail chimera embryos: A test of the quantitative matching hypothesis of neuronal cell death. *J. Neurosci.* 6:2889–2899.

Taniuchi, M., H. B. Clark, and E. M. Johnson, Jr. 1986. Induction of nerve growth factor receptor in Schwann cells after axotomy. *Proc. Natl. Acad. Sci.* (U.S.) 83:4094–4098.

Taniuchi, M., H. B. Clark, J. B. Schweitzer, and E. M. Johnson, Jr. 1988. Expression of nerve growth factor receptors by Schwann cells of axotomized peripheral nerves: Ultrastructural location, suppression by axonal contact, and binding properties. *J. Neurosci.* 8:664–681.

Teissier, G. 1939. Biométrie de la cellule. *Tabulae Biologicae* 19 (pt. 1):1–64.

Tello, J. F. 1917. Génesis de las terminaciones nerviosas motrices y sensitivas. *Trab. Lab. Invest. Biol. Univ. Madr.* 15:101–199.

Terni, T. 1922. Ricerche sulla struttura e sull'evoluzione del simpatico dell'uomo. *Monitore Zool. Ital.* (Firenze) 33:63–72.

Teyler, T. J., and P. DiScenna. 1987. Long-term potentiation. *Ann. Rev. Neurosci.* 10:131–161.

Thoenen, H., and Y.-A. Barde. 1980. Physiology of nerve growth factor. *Physiol. Rev.* 60:1284–1335.

Thoenen, H., Y.-A. Barde, A. M. Davies, and J. E. Johnson. 1987. Neurotrophic factors and neuronal death. *Ciba Found. Symp.* 126:82–95.

Thoenen, H., and D. Edgar. 1985. Neurotrophic factors. *Science* 229:238–242.

Thompson, D. W. 1917. *On Growth and Form*. Cambridge: Cambridge University Press.

Thompson, W. J. 1983. Synapse elimination in neonatal rat muscle is sensitive to pattern of muscle use. *Nature* 302:614–616.

——— 1985. Activity and synapse elimination at the neuromuscular junction. *Cell. and Molec. Neurobiol.* 5:167–182.

Thompson, W. J., D. P. Kuffler, and J. K. S. Jansen. 1979. The effect of prolonged, reversible block of nerve impulses on the elimination of polyneuronal innervation of new-born rat skeletal muscle fibers. *Neurosci.* 4:271–281.

Thorpe, W. H. 1961. *Bird Song: The Biology of Vocal Communication and Expression in Birds.* Cambridge: Cambridge University Press.

Torvik, A. 1956. Transneuronal changes in the inferior olive and pontine nuclei in kittens. *J. Neuropathol. Exp. Neurol.* 15:119–145.

Toulouse, G., S. Dehaene, and J.-P. Changeux. 1986. Spin glass model of learning by selection. *Proc. Natl. Acad. Sci.* (U.S.) 83:1695–1698.

Tower, D. B. 1954. Correlation of neuron density with brain size. *J. Comp. Neurol.* 101:19–51.

Tower, S. S. 1939. The reaction of muscle to denervation. *Physiol. Rev.* 19:1–48.

Townes-Anderson, E., and G. Raviola. 1978. Degeneration and regeneration of autonomic nerve endings in the anterior part of rhesus monkey ciliary muscle. *J. Neurocytol.* 7:583–600.

Truman, J. W., and S. E. Reiss. 1976. Dendritic reorganization of an identified motoneuron during metamorphosis of the tobacco hornworm moth. *Science* 192:477–479.

Tuffery, A. R. 1971. Growth and degeneration of motor endplates in normal cat hind limb muscles. *J. Anat.* 110:221–247.

Usdin, T. B., and G. D. Fischbach. 1986. Purification and characterization of a polypeptide from chick brain that promotes the accumulation of acetylcholine receptors in chick myotubes. *J. Cell Biol.* 103:493–507.

Van der Loos, H., and J. Dörfl. 1978. Does the skin tell the somatosensory cortex how to construct a map of the periphery? *Neurosci. Letters* 7:23–30.

Van der Loos, H., J. Dörfl, and E. Welker. 1984. Variation in pattern of mystacial vibrissae. *J. Heredity* 75:326–336.

Van der Loos, H., and E. Welker. 1985. Development and plasticity of somatosensory brain maps. In *Development, Organization, and Processing in Somatosensory Pathways,* ed. M. Rowe and W. D. Willis, Jr. New York: Alan R. Liss, pp. 53–67.

Van der Loos, H., E. Welker, J. Dörfl, and G. Rumo. 1986. Selective breeding for variations in patterns of mystacial vibrissae of mice. *J. Heredity* 77:66–82.

Van der Loos, H., and T. A. Woolsey. 1973. Somatosensory cortex: Structural alterations following early injury to sense organs. *Science* 129:395–398.

Van Essen, D. C. 1982. Neuromuscular synapse elimination: Structural, functional, and mechanistic aspects. In *Neuronal Development,* ed. N. C. Spitzer. New York: Plenum Press, pp. 333–371.

Von Noorden, G. K. 1980. *Burian and Von Noorden's Binocular Vision and Ocular Motility: Theory and Management of Strabismus.* St. Louis: Mosby.

Voyvodic, J. T. 1986. A general purpose image processing language (IMAGR) facilitates visualizing neuronal structures in fixed tissue and *in vivo. Soc. Neurosci. Abstr.* 12:390.

——— 1987a. Dendritic geometry of sympathetic ganglion cells is regulated by postganglionic target size. *Soc. Neurosci. Abstr.* 13:574.

——— 1987b. Development and regulation of dendrites in the rat superior cervical ganglion. *J. Neurosci.* 7:904–912.

Vrbová, G., T. Gordon, and R. Jones. 1978. *Nerve-Muscle Interaction.* London: Chapman and Hall.

Vrensen, G., D. DeGroot, and J. Nuñes-Cardozo. 1977. Postnatal development of neurons and synapses in the visual and motor cortex of rabbits: A quantitative light and electron microscopic study. *Brain Res. Bull.* 2:405–416.

Wakshull, E., M. I. Johnson, and H. Burton. 1979. Postnatal rat sympathetic neurons in culture. I. A comparison with embryonic neurons. *J. Neurophysiol.* 42:1410–1425.

Wallace, T. L., and E. M. Johnson, Jr. 1987. Partial purification of a parasympathetic neurotrophic factor in pig lung. *Brain Res.* 411:351–363.

Watson, W. E. 1974. Cellular responses to axotomy and to related procedures. *Brit. Med. Bull.* 30:112–115.

Watters, D., and I. A. Hendry. 1985. Purification of a neurotrophic factor for ciliary neurones from chick intraocular tissue using non-denaturing conditions. *Biochem. Intl.* 11:245–253.

Weeks, J. C., and J. W. Truman. 1984. Neural organization of peptide-activated ecdysis behaviors during the metamorphosis of *Manduca sexta.* II. Retention of the proleg motor pattern despite loss of the prolegs at pupation. *J. Comp. Physiol.* 155:423–433.

——— 1985. Independent steroid control of the fates of motoneurons and their muscles during insect metamorphosis. *J. Neurosci.* 5:2290–2300.

Weisblat, D. A., R. T. Sawyer, and G. S. Stent. 1978. Cell lineage analysis by intracellular injection of a tracer enzyme. *Science* 202:1295–1298.

Weiss, G. M., and J. J. Pysh. 1978. Evidence for loss of Purkinje cell dendrites during late development: A morphometric Golgi analysis in the mouse. *Brain Res.* 154:219–230.

Weiss, P. 1924. Die funktion transplantierter amphibienextremitäten. Aufstellung einer resonanztheorie der motorischen nerventätigkeit auf grund abgestimmter endorgane. *Arch. Mikrosk. Anat. Entwicklungsmech.* 102:635–672.

——— 1928. Erregungspecifität und erregungsresonanz. *Ergebn. Biol.* 3:1–115.

——— 1936. Selectivity controlling the central-peripheral relations in the nervous system. *Biol. Rev.* 11:494–531.

——— 1941. Self-differentiation of the basic patterns of coordination. *Comp. Psychol. Monogr.* 17:1–96.

——— 1942. Lid closure reflex from eyes transplanted to atypical locations in *Triturus torosus:* Evidence of a peripheral origin of sensory specificity. *J. Comp. Neurol.* 77:131–169.

——— 1965. Specificity in the neurosciences. *Neurosci. Res. Prog. Bull.* 3(5):5–36.

——— 1968. Research in retrospect. In *Reflections on Biologic Research,* ed. G. Gabbiani. St. Louis: Warren H. Green, pp. 237–244.

Welker, E., and H. Van der Loos. 1986. Quantitative correlation between barrel-field size and the sensory innervation of the whiskerpad: A comparative study in six strains of mice bred for different patterns of mystacial vibrissae. *J. Neurosci.* 6:3355–3373.

Wells, M. J. 1978. *Octopus: Physiology and Behavior of an Advanced Invertebrate.* London: Chapman and Hall.

Werle, M. J., and A. A. Herrera. 1987. Synaptic competition and the persistence of polyneuronal innervation at frog neuromuscular junctions. *J. Neurobiol.* 18:375–389.

Wernig, A., A. P. Anzil, and A. Bieser. 1981. Light and electron microscopic identification of a nerve sprout in muscle of normal adult frog. *Neurosci. Letters* 21:261–266.

Wernig, A., and M. Fischer. 1986. The nerve-muscle junction: A remodelling contact. *Exp. Brain Res.* (suppl. 14):245–255.

Wernig, A., and A. A. Herrera. 1986. Sprouting and remodelling at the nerve-muscle junction. *Prog. in Neurobiol.* 27:251–291.

Wernig, A., M. Pécot-Dechavassine, and H. Stöver. 1980a. Sprouting and regression of the nerve at the frog neuromuscular junction in normal conditions and after prolonged paralysis with curare. *J. Neurocytol.* 9:277–303.

——— 1980b. Signs of nerve regression and sprouting in the frog neuromuscular synapse. In *Ontogenesis and Functional Mechanisms of Peripheral Synapses,* INSERM Symp. 13, ed. J. Taxi. Amsterdam: Elsevier, pp. 225–238.

Westerfield, M., J. V. McMurray, and J. S. Eisen. 1986. Identified motoneurons and their innervation of axial muscles in the zebrafish. *J. Neurosci.* 6:2267–2277.

White, J. G., E. Southgate, J. N. Thomson, and S. Brenner. 1976. The structure of the ventral nerve cord of *Caenorhabditis elegans. Phil. Trans. R. Soc. Lond.,* ser. B, 275:327–348.

Whitington, P. M., M. Bate, E. Seifert, K. Ridge, and C. S. Goodman. 1982. Survival and differentiation of identified embryonic neurons in the absence of their target muscles. *Science* 215:973–975.

Whitington, P. M., and E. Seifert. 1982. Axon growth from limb motoneurons in the locust embryo: The effect of target limb removal on the path taken out of the central nervous system. *Dev. Biol.* 93:206–215.

Wiersma, C. A. G. 1931. An experiment on the "resonance theory" of muscular activity. *Arch. Néerl. de Physiol.* 16:337–345.

Wieschaus, E., and W. Gehring. 1976. Clonal analysis of primordial disc cells in the early embryo of *Drosophila melanogaster. Dev. Biol.* 50:249–263.

Wiesel, T. N. 1982. Postnatal development of the visual cortex and the influence of environment. *Nature* 299:583–591.

Wiesel, T. N., and D. H. Hubel. 1963a. Effects of visual deprivation on morphology and physiology of cells in the cat's lateral geniculate body. *J. Neurophysiol.* 26:978–993.

——— 1963b. Single-cell responses in striate cortex of kittens deprived of vision in one eye. *J. Neurophysiol.* 26:1003–1017.

——— 1965. Comparison of the effects of unilateral and bilateral eye closure in cortical unit responses in kittens. *J. Neurophysiol.* 28:1029–1040.

Wigston, D. J. 1980. Suppression of sprouted synapses in axolotl muscle by implanted foreign nerves. *J. Physiol.* (Lond.) 307:355–366.

——— 1987. Repeated imaging of neuromuscular junctions in mouse soleus muscles *in vivo. Soc. Neurosci. Abstr.* 13:1007.

Wilder, B. G. 1911. Exhibition of, and preliminary note upon, a brain of about one-half the average size from a white man of ordinary weight and intelligence. *J. Nerv. Ment. Dis.* 30:95–97.

Williams, L. R., S. Varon, G. M. Peterson, K. Wictorin, W. Fischer, A. Björklund, and F. H. Gage. 1986. Continuous infusion of nerve growth factor prevents basal forebrain neuronal death after fimbria fornix transection. *Proc. Natl. Acad. Sci.* (U.S.) 83:9231–9235.

Williams, R. W., M. J. Bastiani, B. Lia, and L. M. Chalupa. 1986. Growth cones, dying axons and developmental fluctuation in the fiber population of the cat's optic nerve. *J. Comp. Neurol.* 246:32–69.

Williams. R. W., and K. Herrup. 1988. The control of neuron number. *Ann. Rev. Neurosci.* 11:423–453.

Williams, S. R. 1902. Changes accompanying the migration of the eye and observations on the tractus opticus and tectum opticum in *Pseudopleuronectes americanus. Bull. Museum Comp. Zool. Harvard College* 40:1–57.

Wilson, E. O. 1980. *Sociobiology.* Cambridge, Mass.: Harvard University Press.

Wilson, P. A., J. Scott, J. Penschow, J. Coghlan, and R. A. Rush. 1986. Identification and quantification of mRNA for nerve growth factor in histological preparations. *Neurosci. Letters* 64:323–329.

Winfield, D. A. 1983. The postnatal development of synapses in the different laminae of the visual cortex in the normal kitten and in kittens with eyelid suture. *Dev. Brain Res.* 9:155–169.

Winklbauer, R., and P. Hausen. 1985a. Development of the lateral line system in *Xenopus laevis:* III. Development of the supraorbital system in triploid embryos and larvae. *J. Embryol. Exp. Morph.* 88:183–192.

——— 1985b. Development of the lateral line system in *Xenopus laevis:* IV. Pattern formation in the supraorbital system. *J. Embryol. Exp. Morph.* 88:193–207.

Woolsey, C. N. 1958. Organization of somatic sensory and motor areas of the cerebral cortex. In *Biological and Biochemical Bases of Behavior,* ed. H. F. Harlow and C. N. Woolsey. Madison, Wis.: University of Wisconsin Press, pp. 63–82.

Woolsey, C. N., W. H. Marshall, and P. Bard. 1942. Representation of cutaneous tactile sensibility in the cerebral cortex of the monkey as indicated by evoked potentials. *Bull. Johns Hopkins Hosp.* 70:399–441.

Woolsey, T. A., D. Durham, R. M. Harris, D. J. Simons, and K. L. Valentino. 1981. Somatosensory development. In *Development of Perception,* vol. 1, ed. R. N. Aslin, J. R. Alberts, and M. R. Petersen. New York: Academic Press, pp. 259–292.

Woolsey, T. A., and H. Van der Loos. 1970. The structural organization of layer IV in the somatosensory region (SI) of mouse cerebral cortex. The description of a cortical field composed of discrete cytoarchitectonic units. *Brain Res.* 17:205–242.

Yamakado, M., and T. Yohro. 1979. Subdivision of mouse vibrissae on an embryological basis, with descriptions of variations in the number and arrangement of sinus hairs and cortical barrels in BALB/c (nu/ + ; nude; nu/nu) and hairless (hr/hr) strains. *Am. J. Anat.*, 155:153–174.

Yamauchi, A., and G. Burnstock. 1969. Post-natal development of smooth muscle cells in the mouse vas deferens. A fine structural study. *J. Anat.* 104:1–15.

Yawo, H. 1987. Changes in the dendritic geometry of mouse superior cervical ganglion cells following postganglionic axotomy. *J. Neurosci.* 7:3703–3711.

Yee, W. C., and A. Pestronk. 1987. Mechanisms of postsynaptic plasticity: Remodeling of the junctional acetylcholine receptor cluster induced by motor nerve terminal outgrowth. *J. Neurosci.* 7:2019–2024.

Yip, J. W. 1986. Specific innervation of neurons in the paravertebral sympathetic ganglia of the chick. *J. Neurosci.* 6:3459–3464.

Yip, H. K., and E. M. Johnson, Jr. 1984. Developing dorsal root ganglion neurons require trophic support from their central processes: Evidence for a role of retrogradely transported nerve growth factor from the central nervous system to the periphery. *Proc. Natl. Acad. Sci.* (U.S.) 81:6245–6249.

Yoon, M. G. 1971. Reorganization of retinotectal projection following surgical operations on the optic tectum in goldfish. *Exp. Neurol.* 33:395–411.

——— 1972. Transposition of the visual projection from the nasal hemiretina onto the foreign rostral zone of the optic tectum in goldfish. *Exp. Neurol.* 37:451–462.

——— 1976. Progress of topographic regulation of the visual projection in the halved optic tectum of adult goldfish. *J. Physiol.* (Lond.) 257:621–643.

Young, J. Z. 1963. The number and sizes of nerve cells in Octopus. *Proc. Zool. Soc. Lond.* 140:229–254.

——— 1979. Learning as a process of selection and amplification. Hughlings Jackson Lecture. *J. Roy. Soc. Med.* 72:801–814.

Zucker, R. S. 1982. Processes underlying one form of synaptic plasticity: Facilitation. In *Conditioning,* ed. C. D. Woody. New York: Plenum Press, pp. 249–264.

——— 1987. The calcium hypothesis and modulation of transmitter release by hyperpolarizing pulses. *Biophys. J.* 52:347–350.

Acknowledgments

I AM GRATEFUL to friends and colleagues who generously gave valuable comments on this book. They include Nigel Daw, Cindy Forehand, David Gottlieb, Viktor Hamburger, Gene Johnson, Anthony LaMantia, Jeff Lichtman, Albert Roos, Josh Sanes, Bill Snider, Gunther Stent, Paul Taghert, Jim Voyvodic, and Bob Wilkinson. I owe a particular debt to Josh Sanes and Jim Voyvodic, who labored through two versions of the manuscript and provided a great deal of useful criticism. I also owe special thanks to Jeff Lichtman for challenging discussion of these topics over the years and for collaboration on many experiments. Finally, I am deeply indebted to Sue Eads for her skillful and patient typing of the manuscript; to Jane Dunford-Shore for her intelligent help on many different aspects of the book; to Vicki Friedman and Laszlo Meszoly for their talented artwork; and to the Grass Foundation for making it possible to spend three summers at the Marine Biological Laboratory in Woods Hole, where the major part of the book was written.

Index

Ablation, cell, 37
Ablation, target: and synaptic rearrangement, 10, 11; and neuronal death, 45, 46, 123–124; and neuronal form, 55, 57; effect on dendrites, 105–107; and nerve growth factor, 123–124
Action potential, 2, 69, 165–166
Allometric equation, 24
Aplysia californica, 167, 169
Auditory system, 29, 30, 31, 119, 144, 145
Axonal inputs: elimination of, 81–83, 86–87, 88–96; aberrant, 92–93, 94; maintenance in maturity, 99–102, 103, 107–120; and nerve growth factor, 138; competition according to sets, 151–158, 160, 174
Axons: defined, 2; compared to dendrites, 2–3; regulation of branching, 59–60; competition between, 83, 86–87; sprouting of, 99–101; and nerve growth factor, 131–132; bipolar, 138, 159; synchronous activity of, 152–153
Axotomy: postganglionic, 98–99, 101–105, 106, 132; effect on dendrites, 102, 104, 105; neuron survival after, 139

Bat, 29, 30, 31
Brain: evoked potentials in, 18; weight, 20, 23, 24; growth of, 20, 60–61; synaptic rearrangement in, 30–31, 87–96, 118–120; language area in, 42; somatosensory cortex in, 42, 43, 119; cerebral cortex in, 61, 64, 65; neuronal packing density in, 64, 65; visual cortex in, 77, 89–92, 93, 144; corpus callosum in, 92; nerve growth factor in, 125; other trophic factors in, 136–137
Brain size: in development, 20; versus body size, 20, 24; variation between species, 24; variation within species, 41–42; and mental performance, 42; and dendrites, 60, 61
Brain-derived neurotrophic factor, 136–137
Broca, P., 42
Bueker, E., 124

Caenorhabditis elegans, 35, 37, 38
Cat, 68, 77, 89–92, 144, 152
Cell: division, 20, 21, 22; number, and body size, 22–23; lineage, 37, 39–41; number, variation in, 38–42
Cell adhesion molecules, 140
Cell size: and body growth, 20; in muscle, 20; and body size, 22–23; constancy between species, 22–23; in invertebrates, 23. *See also* Neuronal size
Chemoaffinity hypothesis, 4, 6, 7, 160; revival by R.W. Sperry, 8–10; qualifications of, 10
Chick, 45, 46, 48, 49, 88, 124, 129, 137, 144–145, 146
Competition: and neuronal form, 15; and neuronal number, 15, 47; for trophic factors, 15, 47, 49, 134–136, 137, 139, 140–141, 173; and input elimination, 81–83, 86–87; effect of dendrites on, 83, 86; in visual system, 89, 92; among sets of inputs, 151–158, 160, 174
Compound eyes, 10–12
Consciousness, 1, 170
Convergence, 54, 68; and dendritic geometry, 70–73; functional effects of, 71; and electrical activity, 71–73, 151–153

Cricket, 57
Critical periods, 117–120

Degeneration, neuronal. *See* Neuronal death
Dendrites: defined, 2; compared to axons, 2–3; and neural adjustment, 34; in autonomic ganglia, 56–57, 59, 60, 62–63, 64, 66, 71–73, 82–83, 84–85, 86, 102, 104, 105, 114, 116, 117, 133; of Purkinje cells, 56–57, 60–61, 62–63, 88–89; growth of, 60–64; and synaptic contacts, 67; and neural unit size, 71–73; and convergence, 82–83, 84–85; effect on competition, 83, 86; and postganglionic axotomy, 102, 104, 105, 106; maintenance in maturity, 102, 104–120; and neuron-to-target ratio, 105–107; in superior cervical ganglion, 105, 106; remodeling in maturity, 114–116, 117; hormonal influences on, 119; and nerve growth factor, 132, 133; and neural activity, 144, 145
Development: cell division in 20, 21, 22; cell size in, 20, 22; in invertebrates, 35–38; determinate and indeterminate patterns of, 37–41; aberrant axonal projections in, 92, 93, 94
Divergence, 54; and body size, 67–70; functional consequences of, 69–70; importance of variation in, 70; and neural unit size, 70; in parasympathetic ganglia, 149
Dog, 23, 65

Elephant, 23, 65, 67, 68
Embryogenesis, 35–38
Endplate, muscle, 60, 79–81, 108
Evoked potentials, 18
Evolution. *See* Phylogeny

Facilitation, 165–166
Flatfishes, 27, 28
Frog, 9, 27, 28, 81, 110, 114

Ganglia, autonomic: sympathetic chain, 4, 5, 12; variation within species, 41; neuronal death in, 45, 47; neuronal form in, 56–57, 59, 60, 62–63, 64, 66, 71–73, 82–83, 84–85, 86, 102, 104, 105, 114, 116, 117, 133; divergence in, 69–70; electrical activity in, 71–73; convergence in, 71–73, 82–83, 84–85; input loss in, 79, 80; and postganglionic axotomy, 101–105; synaptic rearrangement in, 114–116; nerve growth factor sensitivity of, 125, 126–127, 128–130. *See also* Superior cervical ganglion
Giraffe, 27
Glial cells, 20, 44, 138
Grasshooper, 58
Growth: through cell addition, 20–22; of body versus nervous system, 20, 22; through cell enlargement, 20, 22
Guinea pig, 66, 69, 72–73, 99, 147

Hamburger, V., 123–124
Hamster, 66, 69, 72–73
Hebb, D. O., 155–156
Heterotis niloticus, 31
Hormones, 2, 49, 50, 57, 119–120
Horse, 65

Induction, 3, 39
Innervation, inhibitory, 158–159
Invertebrates: and the trophic theory, 1, 161–165; cell size in, 23; small, development in, 35–38; large, development in, 38; size variation in, 38; neuronal death in, 52; neuronal form in, 55, 57, 58; learning in, 167

Katz, B., 165

Langley, J. N., 4, 6, 8
Lateral geniculate, 89
Learning: and neural plasticity, 121–122; and "Hebbian synapses", 155; and the trophic theory, 165–169, 174
Leech, 57
Levi-Montalcini, R., 123–125
Lobster, 162–164

Maps, neural, 8, 17–19, 97
Mauthner cells, 55
Memory, 165–169
Metamorphosis, frog, 27, 28
Monkey, 18, 20–21, 42–43, 61, 68, 76, 77, 89–93, 152. *See also* Primates
Motor units, 68–69, 121
Moth, 58

Mouse, 23, 25–26, 29, 31–33, 56–57, 65, 66, 68, 69, 72–73, 110, 111–113, 115–116, 126, 137
Muscle: cell size in, 20, 22; growth of, 20, 59–60; synaptogenesis in, 75–76; synaptic rearrangement in, 107–114
Muscle activity, 143; and neuronal death, 146; and input elimination, 147; and axonal sprouting, 147–148; and nerve growth factor availability, 148
Muscle innervation, 59–60, 70, 77–81; of smooth muscle, 3–4, 5, 68–69; and resonance hypothesis, 7–8; and myotypic specification, 8; in frog metamorphosis, 27, 28; polyneuronal, 77–81
Myotypic specification, 8

Nematode, 35, 37, 38
Nerve growth factor, 16; discovery of, 123–124; and neuronal death, 123–124, 125, 126, 132; amino acid sequence of, 124; sources of, 124–125; administration of, 125; specificity of, 125–126; antiserum to, 125, 132; synthesis of, 127, 137–138, 147–148; uptake of, 127; effect on neurites, 128–132; effect on preganglionic innervation, 132–133
Neural activity: and resonance hypothesis, 8; effect on targets, 15, 142–144, 159–160; and trophic factors, 15, 144, 145, 146, 156–158, 174; and convergence, 151–153; synchronous versus asynchronous, 151–156; and synaptic efficacy, 165–166
Neural Darwinism, 172
Neural injury, 4, 138, 174. *See also* Axotomy
Neural plasticity, 4; in maturity, 15, 97–122; and critical periods, 118; in visual system, 118; in brain, 118–120; and injury, 121; and learning, 121–122
Neural units, 69–70, 71
Neuroblasts, 20, 21
Neuromuscular junction, 79–81, 109–114
Neuronal death: trophic basis for, 15, 47–52; and neuronal number, 44–51; in development, 45–49, 124, 125; in autonomic ganglia, 45, 47; and competition, 47; timing of, 47; and neuron-to-target ratio, 47–49; target-derived influences on, 47–52; and afferent innervation, 51–52; in invertebrates, 52; in maturity, 52, 98–99; and aberrant projections, 92–93, 94; and nerve growth factor, 123–124, 125, 126, 132
Neuronal form: determinants of, 15, 55, 57, 59–61, 64, 67, 73–74, 105–107; variation in, 54–55, 56, 57, 59, 73; in invertebrates, 55, 57, 58, 73; and target ablation, 55, 57, 105; hormonal influences on, 57; and muscle innervation, 59–60; and phylogeny, 64–67; and animal size, 64–67, 72, 73; and divergence, 67–70; and convergence, 81–87; and input number, 81–87. *See also* Axons; Dendrites
Neuronal number: influence of targets on, 15, 44–45, 47–49; in development, 20, 21, 22, 34; in spinal cord, 25–26; in superior cervical ganglion, 25–26; and body size, 25–26, 34; in phylogeny, 25–26, 34, 52; in nematode, 35, 38; in octopus, 38; determinants of, 44–51; and neuronal death, 49; initial excess, 49, 51, 169–172; and input elimination, 81–83, 86–87
Neuronal processes. *See* Axons; Dendrites
Neuronal remodeling, 107–117. See also Neural plasticity; Synaptic rearrangement
Neurons: uniqueness of various, 10, 55, 58; growth of, 20; proliferation of, 20, 21, 22, 44; size of, 20, 22, 54–55; variation in size of, 54–55; Retzius, 57; packing density of, 64; cell body size of, 64, 65
Neurotransmitters, 2, 137
Newt, 9

Octopus, 38
Ocular dominance columns, 89–92
Olfactory system, 30–31, 61, 144
Ontogeny. *See* Development
Owl, 29, 30, 31, 152–153

Phylogeny: cell size in, 22–23; and neuronal number, 25–26, 34; and neural adjustment, 29, 31–34; and neuronal form, 64–67; and trophic interactions, 169–172, 174
Plasticity. *See* Neural plasticity

Poliomyelitis, 121
Potentiation, post-tetanic, 166
Primates, 25, 77. *See also* Monkey
Purkinje cells, 56–57, 60–61, 62–63, 88–89

Rabbit, 25–26, 59, 66, 68, 69, 72–73, 82–83, 84–85, 86, 150
Rat, 29, 31, 49–50, 60–61, 62–63, 65–66, 68, 69, 72–73, 76–77, 78, 88, 92–93, 94, 105, 106–107
Receptors, postsynaptic, 156
Redfern, P. A., 77
Regeneration, neural: and chemoaffinity hypothesis, 6, 7; of cervical sympathetic trunk, 6, 7, 12; of optic nerve, 8–10; after postganglionic axotomy, 102, 103, 104
Regulation, 37, 39
Residual calcium hypothesis, 166
Resonance hypothesis, 6–8
Retinotectal system, 9–10, 11
Reversal potential, 156

Seizures, epileptic, 17–18
Shrew, 20, 22, 23
Size principle, 70
Size, body: and neural connections, 1; and neural adjustment, 17, 22, 26, 33–34; in development, 19; variation between species, 19–20, 22; and cell size, 20, 22–23; versus size of nervous system, 20, 22, 23–26, 34; and neuronal number, 25–26; and superior cervical ganglion cells, 25–26; and cell body size, 64, 65; and neuronal form, 64–67; and divergence, 67–70; and motor unit size, 68–69; and neural unit size, 69–70; and synaptogenesis, 76–77
Snake, 81
Songbirds, 119
Sperry, R. W., 8–10, 14, 165
Spinal neurons, 25–26
Sprouting, 99–101, 102, 121, 131, 143, 144, 147
Stent, G., 156
Stereopsis, 29
Superior cervical ganglion: targets of, 4; innervation of, 4, 5, 6, 7; denervation of 6, 7, 12–14; reinnervation of, 6, 7, 12–14; neuronal number in, 25–26; neuronal form in, 66; divergence in, 69–70; and neuron-to-target ratio, 105–107
Synapses, 3; in smooth muscle, 3–4; maintenance in maturity, 15, 172; and neural adjustment, 34; number of, 38; and body size, 76–77; at neuromuscular junction, 79–81; and nerve growth factor, 133–134; "Hebbian," 155
Synaptic loss: in maturity, 101–102; after postganglionic axotomy, 101–102, 103; in muscle, 108–114; net, 170–171
Synaptic rearrangement: in maturity, 15, 97–122; in frog metamorphosis, 27–28; in visual system, 27–28, 29; in muscle, 27–28, 107–114; in peripheral nervous system, 77–81, 95–96; in central nervous system, 87–96; on neuronal surfaces, 61, 114–116; in lobster, 162–164
Synaptogenesis, 3; in superior cervical ganglion, 6, 7, 12–14; duration of, 75–76, 77; and animal size, 75–77; and input loss, 79, 81; after partial denervation, 99–101, 102, 104; in maturity, 100–101, 107–120, 147–148, 172; in muscle, 108–114; after polio, 121; molecules relevant to, 139–140; and cell adhesion molecules, 140

Targets, neural, 3; and resonance hypothesis, 7–8; and myotopic specification, 8; influence on neuronal number, 44–45, 46, 47–50, 52–53; size of, and neuronal death, 47–49, 51; influence on neuronal form, 55, 57, 58; influence in maturity, 98–107; and neural activity, 142–144, 159–160
Toad, 39
Transplantation, limb, 7–8
Trophic factors, 2; and neuronal death, 15, 47, 49; competition for, 15, 47, 49, 134–136, 137, 139, 140–141, 173; and neural activity, 15, 144, 145, 146, 156–158, 174; criteria for, 127–128; scheme of action, 134–136, 174; besides nerve growth factor, 136–137; availability of, 137, 141, 148, 173; diffusibility of, 139, 173. *See also* Nerve growth factor
Trophic interactions, 2; in maturity, 15, 120–122; and neuronal death, 47–52; and neuronal form, 61, 64; and post-

ganglionic axotomy, 102; and electrical activity, 142–160
Trophic theory: statement of, 1, 4, 173; and invertebrates, 1, 161–165; application to various neurons, 3; evidence for, 3, 174; and molecular mechanisms, 15–16; and learning, 165–169; and regressive theories of connectivity, 169–172

Vestibulo-ocular system, 28
Visual system: development in flatfish, 27–28; synaptic rearrangement in, 27–28, 29, 89–92, 115, 118–119; development in mammals, 89–92, 152; critical periods in, 118; effects of neural activity on, 151–155

Weiss, P. A., 6–8, 14
Wernicke, C., 42
Whale, 19–20, 22, 23, 24, 52, 53, 67–68

Zebrafish, 39–41